Other titles in the series

Series A

Published

Volume 1 (1976)

M.F. Greaves (London), Cell Surface Receptors: A Biological Perspective

F. Macfalane Burnet (Melbourne), The Evolution of Receptors and Recognition in the Immune System

K. Resch (Heidelburg), Membrane Associated Events in Lymphocyte Activation

K.N. Brown (London), Specificity in Host-Parasite Interaction

Volume 2 (1976)

D. Givol (Jerusalem), A Structural Basis for Molecular Recognition: The Antibody Case

B.D. Gomperts (London), Calcium and Cell Activation

M.A.B. de Sousa (New York), Cell Traffic

D. Lewis (London), Incompatibility in Flowering Plants

A. Levitski (Jerusalem), Catecholamine Receptors

Volume 3 (1977)

J. Lindstrom (Salk, California), Antibodies to Receptors for Acetylcholine and other Hormones

M. Crandall (Kentucky), Mating-type Interaction in Micro-organisms

H. Furthmayr (New Haven), Erythrocyte Membrane Proteins

M. Silverman (Toronto), Specificity of Membrane Transport

In preparation

Volume 5 (1978)

G. Ashwell (NIH, Bethesda), Hepatic Degradation of Circulating Glycoproteins

K. Weber (Gottingen), Actin

P.A. Lehman (Mexico), Stereoselectivity in Receptor Recognition

S. Razin and S. Rottem (Jerusalem), Reconstitution of Biological Membranes

A.G. Lee (Southhampton), Fluorescence and NMR Studies of Membranes

Series B

Published

The Specificity and Action of Animal, Bacterial and Plant Toxins
edited by P. Cuatrecasas (Burroughs Wellcome, North Carolina)

In preparation

Intercellular Junctions and Synapses in Development
edited by J. Feldman (University of London), N.B. Gilula (Rockefeller University, New York) and J.D. Pitts (University of Glasgow)

Microbial Interactions
edited by J. Reissig (Long Island University, New York)

4 Receptors

Edited by

P. Cuatrecasas

Wellcome Research Laboratory, Research Triangle Park, North Carolina

and

M. F. Greaves

ICRF Tumour Immunology Unit, University College London

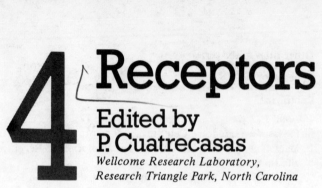

and Recognition

Series A

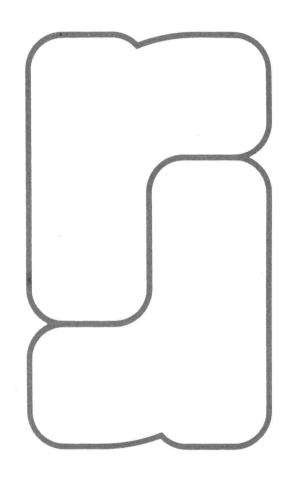

LONDON
CHAPMAN AND HALL

A Halsted Press Book
John Wiley & Sons, New York

First published 1977
by Chapman and Hall Ltd
11 New Fetter Lane, London EC4P 4EE

Typeset by C. Josée Utteridge of Red Lion Setters
and printed in Great Britain at the University Printing House, Cambridge

ISBN 0 412 14330 5 (cased edition)
ISBN 0 412 14340 2 (paperback edition)

Distributed in the U.S.A. by Halsted Press, a Division of John Wiley & Sons, Inc., New York

Library of Congress Cataloging in Publication Data

Main entry under title:

Receptors and recognition, (series A)

Includes bibliographical references.
1. Cellular recognition. 2. Cell receptors.
3. Immunospecificity. I. Cuatrecasas, P. II. Greaves, Melvyn F.
QR182.R4 574.8'761 75–44163
ISBN 0–470–99200–X (v.4)

Contents

Preface *page* vii

1 **Hormone Action at the Plasma Membrane: Biophysical Approaches** 1
M. Sonenberg and A.S. Schneider, Sloan Kettering Institute for Cancer Research and Cornell University, Graduate School of Medical Sciences, New York

2 **The Cellular Receptor for IgE** 75
Henry Metzger, National Institute of Arthritis, Metabolism and Digestive Diseases, National Institutes of Health, Bethesda, Maryland

3 **Endocytosis** 103
Thomas P. Stossel, Medical Oncology Unit, Massachusetts General Hospital and The Department of Medicine, Harvard Medical School, Boston, Massachusetts

4 **Virus Receptors** 141
A. Meager and R.C. Hughes, Division of Biological Sciences, University of Warwick, Coventry and The National Institute for Medical Research, Mill Hill, London

5 **Acetylcholine Receptors** 197
M.E. Eldefrawi and A.T. Eldefrawi, Department of Pharmacology and Experimental Therapeutics, University of Maryland, Baltimore

Preface

As in previous volumes in this series this issue contains a broad spectrum of reviews on cell membrane receptors. Sonenberg and Schneider's article on hormone interaction with cell surfaces provides a comprehensive and critical analysis of current views on hormone–receptor interaction and membrane changes initiated by this union. They present a new integrated model of hormone action which draws a close analogy with the mode of action of neurotransmitter ligands. Ion translocation and alterations in membrane potential are suggested to have a crucial primary role.

Metzger reviews the substantial progress that has been made over the last two years in the study of receptors on mast cells for IgE.

Mast cells have receptors for a particular domain on the Fc portion of IgE molecules which in turn serve as receptors for antigens ('allergens'). This 'double receptor' system has a number of advantages for inquiry into membrane signal transduction events; in particular, the number and specificity of receptors (IgE) for antigens can be experimentally altered, the response-elicited–histamine release–is rapid (few minutes) and rodent leukaemic cells are available as a source of receptors. The leukaemic rat mast cell is defective in triggering through its IgE receptor; this in one sense may limit the usefulness of the system, but it may also shed important light on the normal sequence of causal events in the induction of histamine release. Detailed studies on binding kinetics and preliminary data on the solubilisation and purification of the IgE receptor are reported. The critical membrane events involved in allergen-induced histamine release via Fc_{ϵ} receptor-bound IgE antibodies have not been determined; however it seems likely that calcium ion influx and/or receptor redistribution plays a major initiating or 'messenger' role (see also *Receptors and Recognition* A2, Chapter 2, B.D. Gomperts).

Stossel has reviewed the topic of cell surface membrane events involved in endocytosis. This process is usually studied in the context of uptake and disposal by vertebrate phagocytes (granulocytes, macrophages) of potentially harmful bacteria and viruses. Endocytosis is, however, a property expressed by an enormous diversity of cell types including some from primitive invertebrates. Many of these cells have no involvement in 'protective' responses against microbial or otherwise undesirable intruders. Auto-phagocytosis for example, may be an important clearance operation in certain developmental processes (e.g. tadpole metamorphosis). It seems likely that a considerable diversity of 'recognition' sites are involved in the binding of particulate matter destined for endocytosis. On a single cell, one would expect to find a variety of 'receptors' with varying degrees of specificity. The best studied of these are the opsonin receptors on macrophages. These bear a resemblance to the mast cell receptors discussed by Metzger in that they have a specificity for the Fc of antibody molecules; they differ from the latter, however, in having a distinctive selectivity for certain subclasses of IgG and only bind with high affinity when IgG antibody is complexed (i.e. has 'opsonised') with its antigen. Receptors for complement components (particularly C3) or 'immune-adherence' receptors are also present on macrophages (and other 'myeloid' cells) and are independent physically and functionally from the IgG receptors. Stossel suggests that, not only are there spectra of different independent receptors for endocytosis, but that the membrane responses initiated by these structures are topographically compartmentalised or restricted. In contrast to other receptor-transduction systems, such as those discussed by Sonenberg and Schneider, the initiating ligand does not alter the whole cell surface membrane behaviour, but rather promotes a highly localised re-organisation of that area of membrane to which it has bound. The nature of this response is still unknown but Stossel suggests that focal gelation (or 'contraction') of actin, perhaps promoted by calcium ions, is of particular importance. The process of endocytosis can certainly be modulated by cyclic nucleotides but the mode of action and significance of these effects is still uncertain.

Meager and Hughes provide an extensive discussion of receptors or as they prefer to call them, 'recognition sites' for eukaryotic cell viruses. This is a topic of considerable current interest, although our knowledge in this field is much less advanced than, for example, that of bacterial receptors for phages where serotype selectivity and genetic experiments confer significant practical advantages. As with phagocytosis of bacteria or particles it seems that the uptake of viruses (often by a process resembling phagocytosis) involves a great diversity of recognition structures with considerable variation in specificity. Some viruses, e.g. adenoviruses, pox viruses, have a broad host cell and species range and presumably interact with fairly common membrane components. Others show a much more pronounced specificity (e.g. herpes DNA viruses). An important new concept in this field is that of virus receptor 'families' – introduced by Phillipson and Lonberg-Holm. This idea suggests that distinct receptors exist for groups of viruses which need not necessarily in themselves be related. Unlike studies of soluble ligand (e.g. hormone)–cell interaction, virus receptor studies embody the complexities of cell-cell type interactions. It is therefore more difficult to analyse the specificity of interaction at a chemical level. Considerable advances are being made, however, in the identification of those components of the virus and of the target cell which are involved in the specific interaction. Meager and Hughes emphasise the roles of glycoprotein components and stress the likely importance of oligosaccharide sequences. For example the carbohydrate portion of the glycoprotein 'spikes' on VSV seems to be essential for binding to cells and on the 'other side' of the interaction the haemagglutinating glycoprotein components of the myxoviruses envelope seems to recognise sialic acid terminating oligosaccharide sequences of cell surface proteins and possibly also of lipids (i.e. gangliosides). So far, no virus receptor or 'recognition site' has been isolated, however, some progress has been made in mapping the chromosomal location of structural or regulatory genes for these molecules. We can presume perhaps, that the membrane structures involved have a normal physiological role, which is quite independent of their capacity to recognise (or perhaps be recognised by!) viruses. The latter interaction is, in a sense, a fortuitous complementarity presumably resulting from appropriate evolution, adaptation and selection of invasive micro-organisms (or animal parasites; see *Receptors and Recognition* A1, Chapter 4, K.N. Brown). This provides us with a nice example of the 'selfish gene' concept which is currently exciting ethologists. It is, however, of interest to consider whether the development of these interactions might have taken advantage of membrane sites which serve a physiological initiator role in the endocytotic-phagocytic processes discussed by Stossel. There is one hint that this might be the case; this comes from recent studies indicating a close association and perhaps identity between receptors for Epstein–Barr viruses and complement receptors on human B lymphocytes (see M. Jondal *et al.*, Scand. J. Immunol., 1976, **5**, 401).

Finally, Eldefrawi and Eldefrawi provide a very detailed account of the receptor for acetylcholine – probably the membrane recognition system in which we are closest to a fairly complete resolution of chemical components and mechanisms. This review provides appropriate complementarity so that on hormone receptors which starts this volume; the two together illustrate the way in which important basic mechanisms in this area are beginning to be revealed (see also *Receptors and Recognition* A3, Chapter 5 by Levitski on β-adrenergic receptors – another receptor system in which there has been rapid progress in the past year or two).

February 1977

P. Cuatrecasas
M.F. Greaves

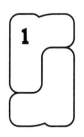

1 Hormone Action at the Plasma Membrane

Biophysical Approaches

MARTIN SONENBERG
and
ALLAN S. SCHNEIDER

Sloan-Kettering Institute for Cancer Research
and
Cornell University, Graduate School of Medical Sciences,
New York

Acknowledgements
We would like to acknowledge the untiring efforts of Ms. M. Priddle, R. Herz, and S. Whaley in the preparation of the typed manuscript.

Receptors and Recognition, Series A, Volume 4
Edited by P. Cuatrecasas and M.F. Greaves
Published in 1977 by Chapman and Hall, 11 New Fetter Lane, London EC4P 4EE

1.1 INTRODUCTION

1.1.1 Nature of receptors

There are many similarities in composition, organization and, to some extent, function of membranes from different tissues and subcellular sites which permit us to speak of 'membrane structure and function'. Yet tissues have different structures and the function of these tissues and their membranes differ in significant details. The specificity of metabolic regulation by the interaction of a ligand of unique structure with the plasma membrane of well-differentiated cells suggests that there is a well-defined receptor(s) in the membrane to recognize such a chemical regulator. This receptor may be composed of a single component or several components in a membrane complex. In the latter case the specificity or molecular complementarity may still reside in one component e.g. glycoprotein, glycolipid [20, 55, 141, 142, 228, 230, 231, 276] but its state may be determined by other molecules e.g. phospholipids.

At the outset, we must ask what are the minimum criteria and what type of data can we accept for demonstrating the presence of a putative receptor. Such a receptor should have an affinity to permit binding at physiological concentrations, be saturable and be structurally specific with regard to the ligand. The binding to such a receptor should be physiologically relevant particularly with regard to the specificity shown by many hormones for their target tissues. Since all the cell surface interactions we shall discuss involve non-covalent interactions, one might expect that such binding to a receptor is reversible. However, since there may be other mechanisms for freeing the receptor of its ligand, e.g. degradation, true reversibility would not seem to be a necessary condition for a receptor. In addition, although many specific ligands bind to cell surface receptors very quickly ($k_1 \simeq 10^5 - 10^8$ l mol^{-1} s^{-1}) [164] this too need not be a necessary condition for a receptor.

In most experiments where a receptor has been demonstrated, direct binding studies with a labeled ligand have been performed. In some situations, direct binding studies have not rigorously established the presence of a receptor although biological experiments suggest strongly that a receptor is present. This has been conspicuously the case with non-nucleated and nucleated erythrocytes. Human growth hormone binding to human erythrocytes has not been established, yet there is biological evidence [140, 201, 245, 284] to indicate interaction. Similarly, insulin alters the maximal net glucose flux in human erythrocytes [355] but direct binding has not established a receptor in this tissue although such binding has been demonstrated [118] in turkey erythrocytes. Catecholamines have been reported to affect a variety of biological end points, notably adenylate cyclase in the nucleated avian erythrocyte [184] and deformability [6] and calcium flux in the human non-nucleated erythrocyte [271]. By binding studies, a saturable high affinity receptor for catecholamines has been demonstrated but with no sterospecificity (summarized in [125, 184]).

Some questions have understandably been raised about the biological relevance of experiments done in the absence of a receptor demonstrated by binding studies or with concentrations of hormone which are 'sub-physiological'. Experimentally, it may be most difficult to establish a limited number of specific receptor sites in the presence of a large amount of non-specific binding. It has been suggested [206] that the avian erythrocyte may have as few as 100 specific binding sites for catecholamines and more than 10 000 non-specific sites [191, 193]. In cases where few [191, 193, 206] or no specific sites can be demonstrated by binding experiments, it may be advantageous to employ other end points e.g. spectroscopic, biologic. If indeed, there are a limited number of specific binding sites for peptide hormones [118], catecholamines [206], prostaglandins [179, 180], acetylcholine [149] in the erythrocyte, these may still interact with 'sub-physiological' concentrations of specific ligands. It may be more relevant to consider the number of ligand molecules bound to specific cell membrane receptors rather than molar concentration of specific ligand. At very low ratios of the interacting species, one may then be obliged to reconcile the rate of binding with the rate of change of some spectroscopic end point. In the case of the human erythrocyte and epinephrine [6, 180], and prostaglandin [6, 179, 180] effects have been noted at concentrations of 10^{-11} M or approximately 1 to 100 molecules per cell.

Although 'receptor' most often refers to a cell component which interacts with a structurally unique ligand and leads to a biologically relevant response, even this minimum definition can be qualified. It has been suggested [61] that the binding must show 'absolute specificity' and the affinity and the number of binding sites must be consistent with the biological activity of the ligand and the physiological responses in the intact system. While specific binding has been used as the criterion in defining a receptor, it should be recognized that this is a necessary but not sufficient condition. The biological response must also accompany the binding. It is possible to dissociate in part the recognition or binding portion of a macromolecule from other biologically active portions as has been done with bovine growth hormone [332] and cholera toxin [20]. The binding portion of the hormone may interact with equivalent affinity and at an equivalent number of receptor sites and displace the native hormone from its receptor [332], but may have little biological activity [352].

Although the binding isotherms of bioregulators and analogs and the biological response isotherms of these ligands are frequently similar, in a number of circumstances there is not this proportionate response. As noted above there may be binding of inactivated macromolecules or fragments of a hormone which give no or minimal response. Some binding has a broad pH range whereas the biological response is limited to the physiological range. Energy sources, ions and other factors may be required for some biological responses but not for binding. It should not be surprising that the biological response determined under physiological conditions in body fluids with a variety of other factors present would not parallel binding studies performed in simple better-defined media under laboratory conditions.

Of interest in ligand binding are the kinetics of association and dissociation. Although equilibrium is frequently reached in 30–90 minutes with dissociation constants of the order of 10^{-8} – 10^{-10} M, for protein and catecholamine hormones (summarized in [125, 164]) and the maximum number of binding sites per cell is between 1×10^2 and 1×10^6, it is apparent that the forward reaction i.e. association of ligand at physiological concentrations and membranes is so rapid ($k_1 \simeq 10^5 - 10^8$ $1\ mol^{-1}\ s^{-1}$) [164] that some molecules interact within seconds. Whether these additional few associated molecules are sufficient to trigger off a biological response depends on the system under study. There are observations [233], for example, to suggest that the K_d value (10^{-8} M) for binding of vasopressin to beef kidney plasma membranes is significantly greater than the apparent K_m (3×10^{-10} M) for adenylate cyclase activation. This would suggest that only a small fraction of available receptors need be occupied to obtain biologically meaningful responses.

1.1.2 Cell surface hormone action

Control of cellular activity by specific bio-regulators interacting with the cell surface may be considered to operate in a series of steps. The first step is the binding of the hormone to the specific receptor or receptor – effector complex in the membrane. In the second step there may be either a coupling of the hormone – receptor complex to one or more membrane effectors which translate the hormonal message or the receptor itself may be capable of producing the second message. The third step is the activation of the membrane effector whether it is an enzyme, transport protein, ionophore etc. It is this step which produces a second messenger(s) that leads to an array of biochemical and biological effects associated with a particular hormone.

This broad formulation for the sequence of events associated with ligand interaction with membrane receptors is largely heuristic and demands additional data and answers to specific questions. What is the nature of membrane receptors? Do they all have similar physical and chemical properties? Is it necessary that all receptors contain a macromolecule alone or in association with other components in order to provide the necessary determinants for ligand recognition? Could smaller molecules alone e.g. gangliosides [20, 55, 141, 142, 228, 230, 231, 276] provide such a structure? Are receptors linked to a large array of 'effectors' so that a large number of ligand molecules must be bound for complete biological expression or may a few molecules initiate a concerted or amplified reaction? What structures and reactions are interposed between the ligand–receptor and the biological effectors? After interaction of receptor and ligand, how many of the reactions are reversible and how many irreversible? What are the requirements for reversibility? For the purposes of study, how many of these may be eliminated and yet retain the 'normal' biological response? In other words, is it possible to study meaningfully isolated membrane reactions in the absence of an intact membrane? In this context, how meaningful are the observations being made with model systems, subfractions of membranes, membranes perturbed by probes or chemical reactions?

In the present chapter we propose to examine the physico-chemical events associated with the specific binding of a hormone to the receptor in a target tissue. We will be particularly concerned with those changes which may be involved in the coupling of the hormone–receptor binding with enzyme activation, ion translocation, and/or other biological expression. Although we are most interested in hormones active at the cell surface, it would appear that the changes associated with other ligand binding may have features in common. Thus, experiments with neurotransmitters (acetylcholine, catecholamines) and peptide toxins (cholera toxin and colicin) will also be analyzed.

Initially, we shall examine the mobility and topography of hormone receptors insofar as they may relate to our major concern i.e. the transduction of the hormone–receptor message. Subsequently we shall examine *in situ* physicochemical changes of membrane proteins and lipids reflected in intrinsic and extrinsic spectroscopic properties. Hormone-induced ion translocations will also be examined. Finally, we shall propose an integrated model for the mechanism of hormone action at the cell surface which brings together the data from these three areas with other relevant information on hormone-induced membrane changes e.g. binding regulation, multiple enzyme activation, and the membrane-regulated dynamic steady state of the cell which is shifted by the hormone.

1.2 MOBILITY AND TOPOGRAPHY OF CELL SURFACE RECEPTORS

An excellent review of the structure of biological membranes by Stoekenius and Engleman [321] appeared in 1969 in which evidence for the then current models of membrane structure was critically examined. The authors' conclusions were in favor of the lipid bilayer model as opposed to the lipoprotein subunit model. What is noteworthy about this review from our present perspective is that as recently as 1969 virtually no mention was made of the mobility, topographical variability, and asymmetry of membrane components. In fact, in the comprehensive chapter on membrane structure by Singer published in 1971 [306] the proposed 'lipid-globular protein mosaic' model did not specifically include lateral mobility of membrane components. It contained only a short footnote mentioning the possibility of translational diffusion of proteins and lipids in the plane of the membrane, citing the now much celebrated work of Frye and Edidin [113]. The fluidity and disorder of the phospholipid hydrocarbon chains in the interior of the bilayer were well established by 1970 to be sure. However, fluidity or 'wagging' of the fatty lipid tails in the non-polar membrane interior is clearly not the same phenomenon as lateral diffusion of entire phospholipid molecules in the plane of the membrane. It is this lateral mobility of membrane lipids and proteins and the consequent variable topography of membrane components that has been one of the most important additions to the basic Gorter–Grendel [123]-Danielli–Davson [67] lipid bilayer model in recent times. In addition, in the last few years there have been a variety of very elegant experiments providing insight into (a) the role of microtubules and microfilaments in regulating membrane protein mobility [21, 77, 84, 85, 236, 346–351] (b) the transverse mobility or flip–flop rates of phospholipids across the membrane [162, 176, 220, 286] (c) the inside-outside asymmetry of membrane proteins [317] and lipids [34, 35, 336, 356], and (d) the state and possible functions of membrane-bound water [104, 105, 181, 293] and divalent cations [208, 256, 267, 268, 302, 303]. In this section we will review the recent literature on the mobility and topography of membrane components with special reference to cell surface hormone receptors. We will only briefly touch upon lectin and antigen receptors where relevant to the argument, since this literature has recently been comprehensively reviewed elsewhere [36, 84, 236, 316]. We first outline current ideas on membrane component mobility, fluidity, and topography and then proceed to a discussion of specific receptor systems, emphasizing hormone receptors.

1.2.1 Dynamic aspects of membrane structure

We would like to clarify at the outset several distinct and possibly independent phenomena which have been described loosely in the literature in a sometimes confusing manner. These are the concepts of membrane fluidity, membrane component mobility, and membrane topography. By membrane fluidity, we understand the microviscosity of the lipid phase (or phases) within which membrane protein and lipid molecules diffuse. By lateral mobility of membrane components, we mean the translational motion of individual protein or lipid molecules or aggregates in the plane of the membrane. By transverse mobility we mean the 'flip–flop' motion of lipids and proteins perpendicular to and across the plane of the membrane. The topography of the membrane refers to the position and distribution of components in the membrane at any given time and under a defined set of environmental conditions. Thus, mobility of membrane components could result in an altered topographical distribution of membrane receptor proteins but doesn't have to. Similarly a change in membrane topography would reflect a change in mobility of membrane components but not necessarily an increase in fluidity since lateral mobility of membrane components can occur without changes in membrane viscosity. In fact, fluorescence polarization measurements have demonstrated an increase in membrane fluidity without an accompanying increase in membrane protein mobility [114, 301]. We therefore caution against the use of altered membrane topography, as frequently seen in electron micrographs, as sufficient evidence of increased membrane fluidity.

Mobility, topography and asymmetry of membrane components

The dynamic aspects of membrane structure have recently been reviewed by several authors [36, 84, 86, 236, 316] including the possible role of cytoplasmic regulation of receptor mobility and topography by microtubules and microfilaments. The picture that has emerged is one in which the basic structural element of most biological membranes is a lipid bilayer in which the individual phospholipid and protein molecules can diffuse laterally in the plane of the membrane. The diffusion constant of membrane phospholipids has been measured by electron spin resonance using spin labels [79, 290, 320, 329] and by nuclear magnetic resonance [70, 145, 183] and generally found to be of the order of $10^{-7}-10^{-9}$ cm^2 s^{-1}. The rate of protein lateral diffusion in the membrane plane varies in the range of $D = 10^{-9}-10^{-12}$ cm^2 s^{-1} depending on the particular protein and membrane under consideration [86, 87, 242, 248, 260]. In some membranes the proteins may be essentially immobilized in patches as in the purple membrane of *Halobacterium halobium* [242], and in acetylcholine receptor clusters [50]. From the two dimensional Einstein diffusion equation [88],

$$D = \frac{(\Delta x)^2}{4t},$$

relating the diffusion constant D (cm^2 s^{-1}), to the distance traveled Δx (cm) in a time, t(s), we can determine the lateral velocity of membrane components. Thus membrane lipids with $D \simeq 10^{-9}-10^{-7}$ cm^2 s^{-1} would move laterally a distance of 0.6–6.0 μm in one second, and could range over 10 μm of cell surface in a time of a few seconds to a few minutes. Similarly, membrane proteins with diffusion constants in the range of $10^{-9}-10^{-12}$ cm^2 s^{-1} would laterally diffuse 0.02–0.6 μm in one second. Depending on the protein and the membrane, it might take anywhere from about 4 minutes to 3 days for a membrane protein to travel say 10 μm laterally over a cell surface. Insofar as lateral diffusion is occuring in the viscous medium of the bilayer one might expect an inverse linear relationship between membrane viscosity and the diffusion constants of membrane proteins and lipids.

In addition to lateral diffusion, membrane lipids can undergo transverse 'flip–flop' motion across the membrane. The rates for this type of motion are somewhat slower, with characteristic half-times varying depending on the membrane and particular phospholipid under study. In erythrocyte membranes, a half-time of 20–30 minutes has been reported [220], while in eel electroplax vesicles the half-time was 3–7 minutes. 'Flip–flop' rates in model lipid vesicles were considerably slower with half-times of the order of 6 hours [162, 176] and longer [162, 286].

There have been no reports of membrane proteins moving from inner to outer membrane faces nor in the opposite direction. However, Blasie claims to have measured rhodopsin 'sinking' further into the lipid hydrocarbon core of isolated disc membranes after bleaching [29]. It is also conceivable that a hydrophobic subunit of a peptide ligand could immerse into the lipid bilayer after binding to its receptor as has been postulated for the active subunit of cholera toxin [19, 20].

There have been numerous other measurements reported in the literature which provide qualitative evidence of protein lateral mobility on the basis of changes in the topographical distribution of labeled proteins and intramembranous particles. The latter have been observed by freeze–fracture electron miscroscopy and are assumed to be membrane proteins imbedded in the bilayer [91, 257, 258, 315, 327]. The former were observed in electron micrographs with markers such as ferritin [237] and colloidal iron [235]. Fluorescent antibody measurements have also provided data on membrane protein mobility [87, 113]. While the normal unperturbed protein distribution tends to be randomized over the surface of free cells in suspension, a variety of states of aggregation of membrane proteins into small clusters, patches, and polar-like caps have been observed upon induction by multivalent ligands, e.g. antibodies (for reviews see [36, 84, 236, 316]) as well as by proteolytic enzymes and pH [257, 258]. It is thus possible to influence the two dimensional asymmetry of membrane proteins. Their transverse asymmetry, however, seems to be fixed, an observation consistent with the lack of protein transverse mobility across the membrane. The membrane where this has been most studied is probably that of the human erythrocyte, where the oligosaccharide end of the glycoproteins always faces the external aqueous environment and several proteins such as spectrin are always on the inner membrane surface facing the cytoplasm [317]. Stackpole [316] has recently catalogued those membrane proteins known to face either the inner, outer, or both membrane surfaces.

Lateral and transverse asymmetry also exist in the lipid phase of membranes. Thus phosphatidylserine and phosphatidylethanolamine exist principally on the inner half of the erythrocyte membrane bilayer, while phosphatidylcholine, sphingomyelin, cholesterol, and glycolipids are more concentrated in the outer bilayer [34, 35, 336, 356]. Lateral lipid asymmetry can occur in membranes during induced phase separations in the membrane plane. These have been observed in model lipid membranes containing two or more phospholipids as well as in natural membranes [48]. The lipid phase separation might reflect a thermally induced melting or freezing of certain membrane lipids [48, 157, 181] or a divalent cation, e.g. calcium, sequestering of a charged phospholipid such as phosphatidylserine, out of a lipid mixture [157, 246]. Such lipid phase separations tend to exclude membrane proteins from the solid crystal phase and concentrate them in the fluid phase [158, 174, 175].

Dynamic modulation by microtubules and micofilaments

During the last few years several laboratories have implicated cytoplasmic microfilaments and microtubules as regulators of membrane protein mobility and topography (reviewed in [36, 84, 236, 316]). The type of experiment that has generally been reported is the effect of microtubule and microfilament disrupting agents (e.g. colchicine or cytochalasin B) on ligand-induced clustering, patching, and capping of membrane receptors. Capping of lymphocyte and fibroblast receptors has been intensively studied. Cytochalasin B, a potent disrupter of microfilaments, has been found to inhibit lymphocyte lectin receptor and immunoglobulin capping to varying degrees [36, 76, 77, 84, 85, 196, 236, 262, 325, 350]. Colchicine and vinblastine, both widely used microtubule disrupting agents, have no inhibiting effect on lymphocyte receptor capping and may even enhance it [76, 85, 196, 346]. The combination of cytochalasin B and colchicine completely block immunoglobulin capping and partly reverse preformed caps [77].

A related series of experiments have examined the effects of microtubules and microfilaments on the interaction of concanavalin A (Con A) and cell surface antigens. Yahara and Edelman [84, 85, 346–351] have found that non-saturating amounts of Con A are capable of preventing antibody induced patching and capping of lymphocyte surface antigens, a phenomena they call anchorage modulation. This inhibition of capping is reversed when colchicine or other *Vinca* alkaloids are added [85]. Both the original inhibition and its removal by colchicine are reversible effects. The anchorage modulation is considered to be a propagated phenomena since local application of Con A to one portion of the cell surface inhibits movement over the rest of the cell surface [351]. This capping inhibition is dependent on the multivalency and cross-linking capabilities of Con A [85], which is thought to bind to a variety of membrane glycoprotein receptors [84]. Some of these trans-membrane receptors may then be anchored to the submembranous microtubule system. By cross-linking antigen receptors with other glycoproteins which are anchored to cytoplasmic microtubules, Con A could thus inhibit their lateral mobility and capping ability which would be overcome by microtubule disrupting agents such as colchicine. Rapid polymerization of the microtubules under influence of calcium and/or nucleotides could lead to a propagation of the anchoring system across large areas of the cell surface simulating a co-operative membrane response to a surface active ligand.

Edelman [84] has proposed a 'surface modulating assembly' (SMA) consisting of glycoprotein receptors, microtubules, and microfilaments. This SMA package was suggested to influence not only receptor mobility and topography but also immune recognition, embryonic cell interaction and cell growth [84].

Related models have been proposed for the role of cytoplasmic micro-tubules and microfilaments in transmembrane control of receptor mobility and topography by Berlin and co-workers [21–23, 243, 353], by Stackpole [316], and by Post *et al.* [261, 262]. All of these models recognize the receptor anchoring potential of microtubules and the actin-like contractile capability of submembranous microfilaments in regulating cell surface receptor display.

Membrane-bound water

We have thus far outlined the dynamic properties of membrane lipids and proteins. There is a third major component of biological membranes that has generally been neglected in current models of membrane structure. It may be important in stabilizing the bilayer, in the mobility of membrane proteins and lipids, and in the transport of aqueous solutes across the membrane. This is the membrane-bound water or membrane hydration. Direct binding studies by Schneider and Schneider [293] have determined that about 0.2 g water per g dry erythrocyte membranes is very tightly bound, having binding energies comparable to and greater than water molecules in ice. Finean [105] has shown that this is the minimum amount of water required for membranes to exhibit their characteristic bilayer X-ray patterns and Ladbrooke and Chapman [181] have demonstrated calorimetrically that this bound water does not exhibit the ice–liquid water phase transition. Calorimetry experiments suggested that below the 25% hydration level a separation of cholesterol and phospholipid occurs into different regions of the membrane [181]. No comparable effects of dehydration were found on membrane protein secondary structure as measured by circular dichroism [293]. More recently Finch and Schneider [104] have measured the self-diffusion constant of water bound to erythrocyte membranes using proton magnetic resonance techniques. A value of $D \simeq 10^{-9}$ cm^2 s^{-1} was found for the bound water self-diffusion constant. This is four orders of magnitude slower than the liquid water diffusion constant ($D \simeq 10^{-5}$ cm^2 s^{-1}) and provocatively close to the rates of lateral diffusion of membrane proteins and lipids. Furthermore, a similar value can be estimated for the rates of aqueous diffusional transport across the membrane ($D \simeq 10^{-9}$ cm^2 s^{-1}) from water permeability coefficients [104, 319]. There is thus the possibility that some kind of 'iceberg' or water lattice coats all polar surfaces of the membrane, providing anchoring sites which influence the lateral mobility of membrane proteins and/or lipids and which inhibit lipid and protein 'flip-flop' transverse mobility. The low-mobility bound water could also be a rate-limiting barrier to aqueous diffusional transport across the membrane. Further work is needed in this area to corrobrate these early results and extend our knowledge of the properties of the membrane-bound water component.

The fluid-mosaic model–use and limitations

Although the fluid mosaic model [308] of membrane structure has proved very useful, it has several important limitations. The mosaic concept of membrane structure had been suggested some forty years before Singer and Nicolson by Collander and Bärlund [53] who proposed the 'lipoid–sieve' theory of membrane permeability and pictured the membrane structure to be a mosaic of lipoid areas and sieve-like areas (for a review of early membrane models see [33] and [139]). The lipid bilayer as the key structural element of biomembranes is also quite an old concept dating to the work of Gorter and Grendel [123] and Danielli and Davson [67]. What is relatively new is the dynamic aspect of membrane structure and the membrane's capability of variable topography including multiple receptor configurations. Insofar as the fluid mosaic model brings all these concepts together it serves a useful purpose in providing an instant picture of a dynamic membrane structure.

There are, however, serious shortcomings in the model. It neglects a certain amount of relevant data on membrane structure and reflects our limited knowledge of diverse membrane systems. First it should be mentioned that the great majority of data on membrane structure and mobility derives from freely suspended cells such as erythrocytes and lymphocytes as opposed to immobilized cells in solid tissue. The latter more representative cells contain a variety of types of junctions which are likely to impose limitations on the lateral mobility of receptors and maintain heterogeneities, i.e. polarity, in the cell surfaces facing different tissue environments, such as the luminal vs. peritubular sides of a proximal kidney tubule cell [232]. Thus, in solid tissue, dynamic membrane phenomena such as receptor patching and capping are unlikely to occur on the global scale observed for lymphocytes, but may occur on one face of the cell.

Another limitation of the fluid mosaic model is its neglect of the role of cations in influencing membrane structure. It is known that removal of inorganic cations and particularly calcium from membranes results in their breakdown, fragmentation, and partial solubilization [278, 318]. Membrane cations are also vital to excitable membrane function [324]. Furthermore, in model membrane systems, calcium has been shown to induce phase separation of negatively charged phospholipids and subsequent fusion of membranes [157, 246].

Divalent cations could easily provide the linkage between negative charges on various membrane components and influence membrane structural stability, topography, permeability and other functions. Manery *et al.* [208, 209, 256] have summarized the various effects of calcium on membranes and others [24, 267–269] have elevated calcium to the role of 'second messenger' on a par with cyclic AMP in the mechanism of hormone action at the cell surface.

The role of bound water as a major membrane component in stabilizing the lipid bilayer, in possibly influencing membrane component lateral and transverse mobility, and in aqueous transport has already been discussed above. Singer [307] has proposed a tetrameric protein pore in the membrane providing ~ 10 Å liquid water-filled channel for aqueous transport. It is hard to imagine a channel this small being filled with liquid water since the first layer alone of bound water (~ 3.3 Å) on each surface would leave little space for liquid water. Such a transport channel would have to allow for diffusion properties [104] characteristic of bound water (orders of magnitude slower than liquid water) in order to provide accurate estimates of transport rates. Thus far, the fluid mosaic model, has not considered either the bound water or cation contribution to membrane structure and function and is thus necessarily limited and incomplete.

Other aspects of membrane structure not adequately accounted for by the fluid mosaic model include the maintenance of lipid asymmetry across the membrane, the greater rate of synthesis and insertion of phospholipids into the outer membrane leaflet of micro-organisms than the transverse lipid flip–flop rates should allow, and the synthesis, insertion, and turnover of membrane proteins. In addition, much recent evidence on the role of microtubules and microfilaments in regulating mobility of membrane components (see above) has been published after the model was proposed. Even with these limitations, the model still provides a useful framework for considering membrane structure and suggesting ideas for experiments and mechanisms of function. In the following paragraphs we apply current ideas on the dynamic aspects of membrane structure to specific hormone receptors and related systems.

1.2.2 Mobility and topography of specific receptor systems

The various topographical configurations of membrane proteins and lipids and their dynamic properties will have relevance to the feasibility of potential receptor mechanisms. Any receptor mechanism would have to conform to the known mobility and asymmetry of membrane components. Questions such as whether adenylate cyclase and a hormone receptor are totally separate molecules which only recognize each other after hormone binding [19] may be clarified by analysis of receptor topography. The rates of lateral mobility should indicate whether the time scale of possible ligand-induced receptor clustering is consistent with the kinetics of physicochemical changes, enzyme activation, and the biological response. Similarly, the kinetics of hormone-induced membrane co-operative effects would have to be consistent with the molecular dynamics of membrane lipids and proteins. Thus as further information becomes available on cell surface receptor topography and mobility, it will serve both as a guide and test of possible hormone–receptor transduction mechanisms. We now summarize the relatively few studies that have recently been initiated on mapping specific hormone and related cell surface receptors. The relevance of these studies to various proposed models of hormone action will be discussed in the closing section of this chapter.

Insulin

It has been postulated for some time that insulin as well as other peptide hormone receptors are located primarily and possibly exclusively at the external cell surface membrane. Cuatrecasas [59] attempted to demonstrate the ability of insulin to transduce its biological message across fat cell membranes without entering the cell. He used insulin attached to large agarose beads to confine the insulin interaction to the external cell surface and was able to stimulate metabolic effects. The experiments have been criticized however on the basis of stoichiometry and that insulin dissociates from the agarose bead in a biologically active form during storage and incubation [42, 69, 168]. Quite recently several authors have attempted to directly visualize insulin receptors on fat cell membranes [159, 161] and ghosts [305], on intact adipocytes [160] and on isolated liver plasma membranes [244]. Seiss *et al.* [305] have used ferritin-labeled insulin antibodies to localize insulin bound to fat cell ghosts. They found an inhomogeneous distribution of insulin among membrane vesicles with only 20% of the vesicles being labeled. On those vesicles which were labeled a uniform distribution of insulin on the external membrane surface was observed. A relatively high density of ferritin was observed on the labeled membranes, corresponding to about 200 insulin receptors per square micron of membrane surface, at physiological insulin concentrations.

In a similar study of fat cell membranes, Jarett and Smith [159, 161] directly labeled insulin molecules with ferritin, and used electron microscopy to observe their distribution on the membrane. They found that the insulin lies exclusively on the external membrane surface either as individual ferritin cores or clusters of 2–6 ferritin–insulin molecules. About 75% of the membrane vesicles were labeled by the ferritin–insulin conjugate with the remaining unlabeled vesicles possibly being due to inversion of membrane vesicles to exclude the ferritin–insulin complex or to chance thin sectioning which missed an area of ferritin–insulin binding. Pinocytotic microvesicles were also observed to contain insulin receptors exclusively on the membrane surface originally exposed to the external environment. Appropriate controls were cited to indicate the specificity of binding of the ferritin–insulin complex to the insulin receptor, and that the ferritin–insulin conjugate maintained biological activity.

The question arises as to whether the irregular clusters of insulin receptors are induced by insulin binding (as occurs with lectins and antibody receptor clustering and capping [36, 84, 236, 316] or whether this receptor distribution exists before insulin binding. Jarett and Smith have extended their earlier studies [159] to answer this question [161]. A purified ferritin–insulin conjugate (90% monomers, 10% dimers) was incubated with adipocyte plasma membranes under several conditions known to either inhibit or allow migration of membrane proteins in the plane of the membrane. Thus, membranes previously fixed with paraformaldehyde or membranes incubated at 4°C should not allow mobility of insulin receptors while unfixed membranes incubated at 24°C would permit receptor mobility. Incubation of ferritin–insulin (100 μ units ml^{-1}) with membranes under all of the above conditions resulted in the same topographical distribution of insulin receptors – namely, both singly dispersed and in small irregular clusters of 2–6 insulin receptors. These clusters were therefore not due to migration of the ferritin–insulin receptor complexes, but apparently existed in this array prior to insulin binding. These observations may have implications for models of receptor mechanisms involving co-operativity and receptor mobility (see Section 1.5.1 below).

Jarett and Smith have also visualized ferritin–insulin binding to isolated intact adipocytes by electron microscopy [160]. They have confirmed the exclusive binding of the ferritin–insulin complex to the external plasma membrane surface glycocalyx coat. The irregular distribution of insulin receptors in small clusters is consistent with the isolated plasma membrane studies. However, there is significantly less binding to the intact adipocyte than to the isolated plasma membrane, with the intact adipocyte binding being more in line with ^{125}I-insulin binding studies. The excessive binding to the plasma membrane fractions may be related to membrane damage or receptor exposure during isolation. Jarett and Smith [160] have also demonstrated that ^{125}I-insulin binding is almost exclusively localized in the plasma membrane fraction of adipocytes previously incubated with ^{125}I-labeled insulin.

In a related morphological study, Orci and co-workers [244] have used freeze-etching electron microscopy and ferritin-labeled insulin to map insulin receptors on isolated liver plasma membranes. The liver membrane gave a similar distribution of receptors to that found for fat cell membranes, namely a mixed population of both randomly dispersed single receptors and small clusters of several insulin receptors. Small areas of the cell surface were also found to be free of receptors. The insulin binding sites were monitored using ferritin-labeled insulin and the freeze-etch results were compared with parallel thin section negative staining. Because the thin sections view a very small and random area of membrane, the freeze-etching method was considered more useful for mapping the cell surface insulin receptors and relating them to membrane structural elements such as the intramembranous particles. The latter are ubiquitously found in freeze–fracture membrane surfaces and are generally agreed to be proteins imbedded in the membrane bilayer. When Orci *et al.* [244] viewed both the deep-etched and fracture faces of hepatocyte membranes containing bound insulin–ferritin complexes, they found very similar patterns of clustered ferritin molecules and clustered intramembranous particles. This interesting observation suggests the insulin receptors are localized in regions of intramembranous particles and some linkage may exist between the two.

There has been one other study of insulin binding that may have some bearing on the mobility and topography of the insulin receptors. Van Obberghen *et al.* [333] have reported on the effects of microtubule and microfilament disrupting agents on insulin binding to cultured human lymphocytes. They find that cells preincubated with microfilament disrupters (cytochalasins) had a 32% irreversible decrease in insulin binding which they attribute to a reduced number of insulin receptors as opposed to an altered binding affinity. Microtubule disrupters (vinblastine, colchicine) had no effect, while a microtubule stabilizer (D_2O) decreased the affinity (41%) and increased the dissociation rate of ^{125}I-insulin. Van Obberghen *et al.* [333] suggest a possible role of microfilaments in mediating cell surface receptor exposure and distribution.

In summary then, several studies have confirmed that insulin binding occurs on the external surface of the plasma membrane [159–161, 244, 305, 333] and can affect various metabolic activities without entering the cell [159, 160] The receptors seem to exist both singly and in small clusters of two to about ten receptors which may penetrate into the bilayer and be related to the intramembranous particles observed in freeze fracture. The clustered pattern of receptors would seem to exist prior to insulin binding and this pattern may not be affected by the hormone [161]. Microfilaments and/or microtubules may play some role in insulin receptor exposure, topographical distribution and affinity. However, further work is needed to clarify their role and extend the preliminary data thus far reported [333].

Melanocyte-stimulating hormone

Varga *et al.* [335] have studied the topography of ^{125}I-melanocyte stimulating hormone (MSH) receptors on the surface of cultured mouse melanoma cells using light microscope autoradiography. They found a non-random distribution of MSH receptors with the appearance of large clusters or patches on certain portions of the cell surface. Fixation with paraformaldehyde or glutaraldehyde prior to hormone binding did not change the irregular clustered appearance of MSH receptors, thus suggesting the clustering of receptors had existed in the absence of hormone binding and was not hormone-induced. In about 10% of cells a concentration of clusters occured near the two poles of the nucleus, simulating the patching-capping phenomena observed by other workers in lymphocytes [325] and fibroblasts [78]. The degree of spreading of the MSH receptor clusters over the cell surface was temperature-dependent. At 0°C hormone-cell incubation conditions, the receptor clusters were confined to a small portion of the cell surface while at 37°C the receptors were considerably more spread out. Varga *et al.* [335] also mentioned some preliminary data suggesting the receptors lie over the Golgi apparatus and their action may proceed through channels which include the Golgi.

Varga *et al.* have recently extended their studies of the distribution of MSH receptors on melanoma cells by using a double labeling procedure for both MSH and the Golgi complex [334]. In this study they used fluorescein isothiocyanate labeled hormone (FITC–MSH) and a histochemical test for the characteristic Golgi enzyme, thiamine pyrophosphatase. By this double labeling technique they were able to visualize the MSH receptors and the Golgi region on the same cell simultaneously. The result was that the MSH receptors were located in the same area as that outlined by the Golgi-thiamine pyrophosphatase reaction. When comparing the appearance of FITC–MSH on non-melanized vs. melanized cells it was observed that the hormone was on the cell surface of the former and internalized in the latter. The internalized hormone appeared in small patches in about 20% of the melanized cells. Based on the above and related evidence, Varga *et al.* [334] suggest a mechanism by which the hormone, MSH, induces melanization, in two sequential phases:

(1) an MSH-independent, preparatory phase in which pre-melanosomes are formed in the smooth endoplasmic reticulum adjacent to the Golgi, and a tyrosinase is synthesized in the rough endoplasmic reticulum and then reaches the Golgi. The MSH receptors would be synthesized and appear at the surface membrane in a masked form;

(2) a hormone induction phase including hormone binding, cAMP accumulation, pinocytosis with internalization of bound hormone, and

activation of tyrosinase. There is a delay between cAMP formation and tyrosinase activation which may be related to the formation and delivery of pinocytotic vesicles containing the bound hormone to the active centers of melanization near the Golgi. One of the interesting aspects of this mechanism is the possible functional role of hormone receptors being in a patched or capped topographical distribution leading to pinocytosis of bound hormone and subsequent action inside the cell. It is noteworthy that internalization of surface membrane-bound insulin in pinocytotic vesicles has also been observed [159], although no functional relevance has yet been ascribed to this event.

Luteinising hormone

An immunofluorescent technique has been used to label and visualize the luteinising hormone (LH) receptors on the surface of isolated interstitial cells of the rat testis [148]. About 50% of the cells exhibited marked fluorescence and these were thought to be the Leydig cell population of the testis interstitium. About 10% of the cells stained non-specifically and these were subtracted to obtain the percentage of cells exhibiting specific fluorescence. The percentage of cells exhibiting specific fluorescence dropped markedly after hypophysectomy to less than 10%, for cells not previously incubated with human chorionic gonadotrophin and to about 20% for previously treated cells. The observed topography of LH receptors was that of discrete patches distributed over all areas of the cell surface [148], as opposed to a cap at one pole of the cell. The patched state of the receptors was independent of native hormone concentration as indicated by this topographical receptor distribution remaining after hypophysectomy on those cells showing fluorescence [147]. This is an important point in that it indicates that receptor aggregation and patching is not a function of hormone concentration as suggested by some proposed models of hormone action. It also represents another hormone receptor system where a clustered or patched receptor topography is observed in addition to insulin, MSH and acetylcholine.

Growth hormone

The previously cited study [333] of the effects of microtubule and microfilament inhibitors on insulin receptors was also applied to growth hormone receptors. Van Obberghen *et al.* [333] found similar results for both kinds of receptors, with human growth hormone showing a 62% irreversible decrease in binding to cultured lymphocytes in the presence of cytochalasins A, B, and D. This was ascribed to a reduction in the number of receptor sites at the cell surface, with no change in binding affinity. Anti-microtubule agents had no effects on hormone binding. Based on the cytochalasin results, microfilaments may play some role in modulating the appearance of growth hormone receptors at the cell surface.

Catecholamines

Melamed and co-workers [223] have applied a fluorescent derivative of the β-adrenergic blocker, propanolol, to study the distribution of β-adrenergic receptor sites in rat cerebellum tissue sections. Using fluorescence microscopy, they were able to detect the fluorescent receptor probe in the Purkinje cell layer in the cerebellar cortex. The greatest concentration of receptor sites was localized on the apical dendrites of the Purkinje cells. The distribution pattern of fluorescence at low magnification (x 100) was bead-like. Greater intensity of fluorescence was observed on the apical dendrites than on the Purkinje cell bodies. Appropriate controls with (+) and (–) propranolol demonstrated the stereospecificity of binding of the fluorescent derivative to the β-adrenergic receptor. The magnification and resolution of the methods used were insufficient to give a detailed mapping of receptors on individual cell membranes, although a uniform distribution of receptors over the entire Purkinje cell surface could be ruled out.

Acetylcholine

The acetylcholine (ACh) receptor system is among the best-studied receptors, having been isolated and purified and having a well-established pharmacology of cholinomimetic agonists, cholinergic toxins, and specific nicotinic and muscarinic agents [50]. Furthermore the ACh degradative enzyme, acetylcholinesterase (AChE) is well characterized. During recent years there have been a number of studies aimed at localizing or mapping this receptor in the post-synaptic membrane of the eel electroplax [31, 32], the vertebrate neuromuscular junction [9, 68, 100, 266] and the developing chick embryo retina [337]. Most of these studies have used labeled toxins (α-bungarotoxin, α-najatoxin), which specifically bind irreversibly to the ACh receptors, to determine the concentration and distribution of receptors. Autoradiographic, fluorescent, and peroxidase labeling of the toxins have been used and these have been monitored by both light and electron microscopy. In discussing the specific studies below, the neuroanatomical specialization should be kept in mind. In contrast to the other hormones, the neurotransmitter, ACh, is delivered in high concentration at localized synaptic nerve terminals and not uniformly over the entire target cell surface.

Studies of the neuromuscular junction [9, 68, 100, 266] have determined that the ACh receptors (toxin-binding sites) are confined to the endplate region of the sarcolemma and have a heterogeneous distribution over the junctional folds. Extremely high concentrations of receptors are found at the top of the junctional folds which are adjacent and closest to the presynaptic nerve membrane while the depths of the folds were sparse in receptors [9, 68, 100]. Toxin-binding site concentrations reported for the former area of the junctional folds have been estimated at 30 000 ± 27% sites per square micron of membrane surface for mouse sternomastoid muscle [100]. The lower depths of the folds had relatively few sites, only 4% of the concentration near the tips [100]. A rapid decrease of toxin-binding sites, i.e. ACh receptors, to negligible values occurred as areas of the extrajunctional muscle membrane were observed [9, 68, 100]. In contrast, acetylcholinesterase was found uniformly spread over the junctional fold. In a related study [266] of the neuromuscular junction which did not use labeled toxins, but instead directly examined the ultrastructure of the junctional folds using freeze fracture and conventional thin section electron microscopy, irregular rows of densely packed 110–140 Å particles were observed near the tips of the junctional folds. These were thought to correspond to the ACh receptors and toxin binding sites. An estimate of equivalent toxin binding sites which would correspond to the membrane particles resulted in 21 600 sites per square micron, in reasonable agreement with quantitative autoradiographic studies [100].

Similar results have been found by Changeux and co-workers [31, 32] for the post-synaptic innervated membrane of the eel electroplax, using both immunofluorescent [32] and autoradiographic labelled α-najatoxin [31]. The toxin binding sites were confined to the innervated membrane of the electroplax and were in highest concentration (33 000 μm^{-2}) in sub-synaptic areas of this membrane. Areas of the innervated membrane between synapses had a receptor concentration of about 1% of this value and the non-innervated surface of the electroplax had about 0.1% of this concentration. Acetylcholinesterase has been localized almost exclusively on the innervated membrane of the electroplax both under and between the synapses [16, 30, 115]. Acetylcholine receptors and acetylcholinesterase, respectively, have actually been found on different populations of isolated membrane fragments [51, 82, 272]. With electrophysiological experiments using oubain (Na^+/K^+) ATPase has been established on the non-innervated face of the electroplax [165].

The extraordinarily high concentration of receptor sites in synaptic areas of the target cell surfaces suggest an almost close-packed array of receptor proteins into patches, as observed for the densely packed ~ 110 Å particles in freeze-fracture and thin section electron microscopy. X-ray diffraction studies [83] of *Torpedo* electric organ membrane fragments are consistent with these observations. An ordered structure was observed with a 90 Å center to center distance.

A recent study by Vogel and Nirenberg [337] on developing chick embryo retina used ^{125}I-α-bungarotoxin autoradiography to localize the ACh receptors. They find a high concentration of receptors in the synaptic layers of the retina and low receptor concentrations associated with the axons and nerve bodies. Although the resolution in this study was not as high as those previously cited, the findings were consistent with those on the neuromuscular junction and electroplax: acetylcholine receptors are not uniformly distributed over the target cell membrane but are highly concentrated in areas (patches) below the synapse.

Cholera toxin receptors

We depart from hormone receptors in order to examine a peptide ligand receptor system where a substantial body of information has been accumulated [19] indicating close analogy to many aspects of hormone receptor mechanisms including specific high affinity binding to the cell surface membrane and activation of adenylate cyclase. Furthermore, data obtained on this system have contributed to models for the mechanism of hormone action [20, 156]. The system we now consider is cholera toxin binding and mobility on the lymphocyte cell surface.

Cholera toxin (CT) is known to consist of a binding subunit of molecular weight 60 000 (choleragenoid) and an active (adenylate cyclase activation) subunit of molecular weight 36 000 [19]. The receptor is known to be the G_{M1} ganglioside – a monosialic acid glycolipid. Incubating the isolated G_{M1} ganglioside with cholera toxin will block binding of the toxin to cells and incubating the cells with the choleragenoid subunit will do the same. Binding and G_{M1} content could be increased by introducing exogenously added G_{M1} so that each added ganglioside molecule is able to bind one toxin molecule [142]. After binding to a variety of mammalian cells, there is a lag period of at least 20–30 minutes before measurable activity of CT-stimulated adenylate cyclase begins [19]. A question that has arisen is whether lateral diffusion of CT–receptor complexes may be occurring during this lag time with a subsequent redistribution of receptors into an 'active' form [19].

Both Craig and Cuatrecasas [55] and Revesz and Greaves [276, 277] have independently studied the topographical distribution of cholera toxin receptors on lymphocyte cell surfaces by fluorescence microscopy and have obtained more or less consistent results. They find if lymphocytes are first incubated with the toxin at low temperatures (0–4°C), a uniform ring of fluorescence representing a diffuse distribution of receptors is observed. Then if the temperature is raised to 37°C for about 20 minutes or more, capping of receptors occurs. Craig and Cuatrecasas [55] have shown a parallel time course and temperature dependence for the appearance of the receptor capping and the activation of adenylate cyclase. Furthermore, they have also demonstrated that inhibition of capping with anti-cholera toxin IgG also inhibited enzyme activation. The temperature sensitivity of the redistribution from uniform rings into caps is consistent with viscosity-dependent lateral diffusion in the plane of the membrane. Further evidence for the association of toxin receptor and adenylate cyclase comes from solubilization studies. After binding of labeled toxin, activation of adenylate cyclase and solubilization of the fat cell membrane, a peak of ^{125}I-toxin was associated with adenylate cyclase activity on gel filtration [20]. This was apparent only after a time period allowing receptor aggregation. Direct toxin–receptor–cyclase interaction was further indicated by adsorption of solubilized receptor–enzyme complex to agarose derivatives containing 'active' toxin subunits and immunoprecipitation of toxin-stimulated adenylate cyclase by anti-cholera toxin or anti-36 000 molecular weight subunit [20]. Both groups [55, 276, 277] have found an effect of microtubule and microfilament inhibitors (colchicine, cytochalasin B, etc.) in reducing the observed receptor capping. However, only cytochalasin had an inhibitory effect on cyclase activation by cholera toxin [55]. The above is somewhat puzzling since the G_{M1} ganglioside receptor is not long enough (only 30 Å) to bridge the bilayer from the exterior surface and directly attach to cytoplasmic microtubules or microfilaments. Furthermore the capping phenomenon is trypsin-insensitive [276] implying the absence of an exposed protein component of the receptor. The receptor could, however, be altered after cholera toxin binding or perhaps connect via another membrane protein which is buried in the lipid interior.

The cholera toxin-induced patching and capping of receptors has been explained on the basis of multivalency of the toxin. Thus, in a manner similar to lectin-[124] and antibody-induced [325] lymphocyte capping, it is thought that the ligand is at least bivalent and polymerizes into patches and caps after binding to its receptor under the possible influence of microtubules and microfilaments. Craig and Cuatrecasas [55] have demonstrated the multivalency by showing that cholera toxin-labeled lymphocytes will bind to ganglioside labeled agarose beads and that this binding does not occur with toxin-free lymphocytes or if ganglioside, G_{M1}, is in the media to block the free ends of the lymphocyte-bound toxin.

Quite recently Sedlacek *et al.* [298] have confirmed the redistribution of CT receptors on lymphocyte membranes into caps. They directly labeled the receptor with a fluorescent probe by synthesizing a dansylated ganglioside. This is in contrast to the earlier studies [55, 276, 277] which used labeled CT or fluorescent antibodies to CT. The dansylated ganglioside receptor was first incorporated into the lymphocyte membrane at 37°C and resulted in a pronounced surface fluorescence in all observed lymphocytes. After incubation with the CT-binding subunit at 37°C for 20 minutes, distinct capping of receptors could be seen over the uropod region of the lymphocytes. Colchicine reduced the CT-induced cap formation as did pretreatment of lymphocytes with trypsin. Co-capping of lymphocyte anti-IgG receptors was also observed to occur together with the CT receptors at the same pole of the cell.

The cholera toxin receptor capping phenomenon is interesting from the general standpoint of the dynamic structure of membranes in that it demonstrates aggregation, patching and capping of a membrane glycolipid as opposed to the more familiar membrane protein capping induced by lectins or antibodies. The co-capping of immunoreceptors (anti-IgG) and CT receptors at the same pole of the cell implies co-migration of membrane proteins and lipids rather than independent lipid mobility.

The use of CT binding and subsequent adenylate cyclase activation as a general model for hormone receptor mechanisms may be limited since no alteration in receptor clusters occur after hormone binding, and the clustered state of hormone receptors seems independent of the hormone. Furthermore, most adenylate cyclase-stimulating hormones do not exhibit the lag time between binding and enzyme activation. Finally, the time scale of lateral diffusion of membrane receptor proteins (minutes–hours) is longer than some observed times of cyclase activation of a few seconds [281].

Adenylate cyclase

We include studies on the subcellular localization of adenylate cyclase activity in this review since its topographical distribution relative to hormone receptors is clearly relevant to the mechanism of hormone action. There have been several reports [146, 171, 274] citing the use of electron microscopy to visualize lead precipitates of the phosphates liberated by the action of adenylate cyclase on ATP or on adenylate cyclase-specific substrate, adenylylimidodiphosphate (AMP-PNP). Reik *et al.* [274] studied fixed rat liver cells in response to isoproterenol, glucagon, and sodium fluoride. They found the isoproterenol-sensitive adenylate cyclase located almost exclusively on the surface of the liver parenchymal cells, while the glucagon-sensitive cyclase was found primarily on the reticulo-endothelial cells, which comprise 33% of the liver. Fluoride-sensitive cyclase was found on both types of liver cell surfaces. The observed adenylate cyclase activity was not uniformly distributed over the entire parenchymal cell surface membrane, but was confined to the bile caniculus and to the external surface of the microvilli in continuity with the lumen of the liver sinusoid [274]. No products of enzyme reaction were detected intracellularly. The pyrophosphate reaction product was detected only on the outer surface of the plasma membrane. Since there is biochemical evidence for cyclic AMP production on the cytoplasmic side of this membrane [280] it is conceivable that adenylate cyclase spans both surfaces of the plasma membrane and releases its two products on opposite sides. Howell and Whitfield have performed a similar investigation of the A and B pancreatic secretory cells of the islets of Langerhans [146]. They found a uniform distribution of cyclase activity around and confined to the plasma membrane, with no evidence of adenylate cyclase catalytic products inside the cells. The cells were fixed with 1% glutaraldehyde and the fixed cells were shown to maintain 20–40% of their adenylate cyclase activity including sensitivity to glucagon and fluoride [146]. The liberated lead precipitate of pyrophosphate and PNP again appeared on the *outer* plasma membrane surface. At exocytotic sites of fusion of secretory vesicles with the plasma membrane, a small break in adenylate cyclase localization was detected.

Kim has performed a related set of experiments on mucous and serous secretory cells of rat salivary glands (parotid and sublinguinal) [171]. Pieces of fixed glands were incubated with ATP or adenylylimidodiphosphate (AMP-PNP) as substrate and inorganic pyrophosphate or PNP liberated via adenylate cyclase action was precipitated by lead ions at the sites of production. In both the parotid and sublinguinal glands, adenylate cyclase activity was detected along the myoepithelial cell membranes in contact with the secretory cells. It has been suggested [171] that this may be related to the contractile function of the myoepithelial cells aiding secretory cells to discharge their products. A high level of adenylate cyclase activity was detected on serous cells localized along the portion of the cell membrane which fuses with the secretory granules, and faces the lumen. Kim speculates [171] that the high level of adenylate cyclase activity observed in areas of fusion of plasma and secretory granule membranes may be related to the mechanism of fusion, exocytosis and secretion. In contrast to the serous cells, the mucous cells did not exhibit any significant adenylate cyclase activity and it is interesting that these cells do not secrete by fusion and exocytosis, but rather by discharge of their granules through gaps formed in the apical cell membrane [171]. Similar experiments using lead precipitates or inorganic phosphate to localize adenylate cyclase activity have been reported for rat kidney [289], rat epididymal pads [340] and isolated retinal rod outer segments [339].

The above studies claiming to have determined the cellular distribution of adenylate cyclase activity have recently come into serious question. Lemay and Jarett [186] have demonstrated for fat cell membranes, fat pad capillaries and pancreatic islet homogenates that 10^{-4} M $Pb(NO_3)_2$ completely inihibited adenylate cyclase activity. The work cited above generally used 4×10^{-3} M $Pb(NO_3)_2$. Furthermore, Lemay and Jarett showed that at a concentration of 2×10^{-3} M, $Pb(NO_3)_2$ was capable of inducing the non-enzymatic hydrolysis of adenylylimidodiphosphate (AMP-PNP) with small quantities of cyclic AMP produced. These authors could visualize a precipitate on both sides of plasma membrane vesicles using electron microscopy. They conclude that the lead phosphate precipitate method cannot be used for localization of cyclase activity due to inhibition of the enzyme and artefacts at high lead concentrations. In several of the previously cited studies [146, 171, 280] the lead precipitate was located exclusively on the outside of the plasma membrane, in contrast to the Lemay and Jarett results showing precipitate on both sides of fat cell plasma membranes. In addition, the production of lead precipitate was shown to be responsive to isoproterenol, glucagon, and NaF [146, 171, 280]. Another difference between the Lemay and Jarett study and those claiming localization of adenylate cyclase activity is the latters' use of fixed tissue specimens while the former used freshly isolated cells. Fixation may restrict the movement of reaction product. Although the adenylate cyclase localization studies appear to be in serious doubt, further work may be needed for complete clarification of the questions raised.

1.3 MEMBRANE SPECTROSCOPIC RESPONSES

1.3.1 Intrinsic membrane effects

Approaches used

Interaction of a ligand with its receptor may affect proteins in the membrane in a number of ways, some of which are demonstrable by physico-chemical techniques. There may be a generalized change in state or conformation of many membrane proteins or there may be a modification of one or a few specific proteins. The latter could be reflected in a change in (a) enzyme activity e.g. adenylate cyclase, guanylate cyclase, protein kinase, phosphodiesterase, ATPases, (b) ligand-binding activity, (c) ligand degradation activity, and/or (d) the appearance in a functional group [322, 323] e.g. sulphydryl associated with a particular biological activity.

Spectroscopic techniques have been applied to isolated membrane fragments in an attempt to detect overall changes in state of membrane proteins. Reports since 1965 had suggested that it may be possible to evaluate protein conformation in chloroplasts [169] and red cell membranes [187, 341] by circular dichroism. It was apparent after the initial application of optical rotatory dispersion and circular dichroism to membrane systems that distortions in the spectra of optical activity were introduced as a result of light scattering and absorption flattening of turbid samples [291]. Such spectra have since been examined theoretically and experimentally and appropriate corrections suggested in order to obtain more accurate estimates of membrane protein conformation [119, 291, 292].

Ligand-induced effects

Notwithstanding the difficulty in making precise estimates of membrane protein conformation, circular dichroism has been applied to membrane systems in the absence and presence of a peptide hormone. Human growth hormone has been reported [140, 201, 245, 284] to produce metabolic changes in human erythrocytes *in vitro* although no binding or receptor for human growth hormone by human erythrocytes has yet been established. Human growth hormone has been noted [310] to produce a decrease in optical activity in human erythrocyte membranes. Since there were no associated changes in light scattering, these effects were considered to reflect a change in protein conformation i.e. helicity of membrane proteins. However a differential scattering contribution to the observed circular dichroism changes has not been ruled out. Similar effects were not noted [310] with bovine growth hormone. This is consistent with the fact that human, but not bovine growth hormone is biologically effective in humans. The changes in optical activity were small i.e. approximately 5%. Recent unpublished data (Sonenberg) suggest a smaller hormone-induced change in erythrocyte membrane CD (~ 5%) than previously published [310]. Even small changes in CD spectra, however, reflect average protein conformation and are likely to derive from more than one protein. Five percent abundance would have to have a 100% change in optical activity. This was considered unlikely and it was concluded [310] that the membrane changes reflected smaller changes in more than one membrane protein. Moreover, since effects were demonstrable with as few as 70–100 growth hormone molecules per erythrocyte ghost, it was estimated that as many as 5×10^4 molecules of protein were affected by 1 molecule of hormone, indicative of a concerted or possibly co-operative effect. However, the membrane response was not proportional to the dose of hormone introduced, and it was not possible to demonstrate the familiar sigmoid dose response curve associated with an interacting system classically considered to be co-operative e.g. allosteric enzymes.

Whereas determination by circular dichroism of the intrinsic optical activity of membrane proteins gives some indication of the secondary structure of these proteins, changes in tertiary structure may be monitored by the intrinsic fluorescence of membrane proteins. Spectrofluorimetry has been applied to the human growth hormone – human erythrocyte membrane system [311]. Human growth hormone produced a decrease in intrinsic fluorescence which had a peak at 332 nm without any shift in peak position. However, these studies have not been confirmed.

Whereas human growth hormone but not bovine growth hormone would be expected to have effects on human tissues, the latter hormone is effective in non-human vertebrate organisms. Of special note is the fact that bovine growth hormone does localize in the liver after administration to the intact rodent [216, 312, 313]. Moreover, bovine growth hormone has been shown to bind to isolated rat hepatocytes [81] and crude pregnant rabbit liver membrane preparations [332].

Rat liver membranes were studied spectropolarimetrically and spectrofluorimetrically [2, 263, 288] in the presence and absence of growth hormone. Membranes of hypophysectomized but not normal or treated rats showed approximately a 10% increase in negative ellipticity in the presence of bovine growth hormone (10^{-9} M). At higher concentrations (10^{-7} – 10^{-6} M) there was less negative ellipticity [288]. Membranes of hypophysectomized, normal, and growth hormone-treated rats all showed decreased emission of fluorescence and a shift in emission peak from 333 to 338 nm in the presence of bovine growth hormone [288]. Polarization of excitation of fluorescence was unchanged. There was no response of these membranes to denatured bovine growth hormone or bovine serum albumin. Liver membranes became unresponsive to growth hormone after storage, sonication or phospholipase A_2 treatment suggesting again that intact membrane structure is necessary for the response to the hormone [288]. These observations [2, 263, 288] are of a preliminary nature and remain to be independently corroborated.

The nature of the changes of state in the rat liver membrane protein is not clear. Although the change in fluorescence and circular dichroism of liver membranes of hypophysectomized rats is consistent with a change in protein conformation, of concern is the continued response to growth hormone in the membranes of 'growth hormone sufficient' rats as detected by intrinsic fluorescence changes but not by circular dichroism. The dissociation of rotatory and fluorescence response is not unique and has been demonstrated in isolated globular proteins [247]. The specific proteins undergoing change might only alter the environment of aromatic fluorescent side chains without undergoing significant secondary structure changes. It may also be that intrinsic fluorescence determination is more sensitive and may thus detect changes in a limited number of proteins e.g. receptors. Circular dichroism might require major changes in conformation in degree or number of proteins before it could be detected.

Of additional concern in these studies are the reported [2, 263, 288, 310] changes in the sub-physiological concentration range i.e. ca. 10^{-15} M. Since changes were apparent very rapidly, i.e. less than one minute, it would appear that this would be unlikely in a process with a binding rate constant of 10^5 l mol^{-1} s^{-1}. Of course, in a 'bimolecular' reaction one would also have to take into account the concentration of cells and the number of receptor sites per membrane (possibly 10 000 per cell). However, neither the receptor site nor the membranes can be considered to diffuse freely in a manner similar to the ligand in solution. On the basis of diffusion alone, the association of uncharged molecules should be defined by the Smoluchowski equation [3, 309]

$$k_a = \frac{4\pi N}{1000} r_{12} (D_1 + D_2)$$

where N is Avogadro's number, D_1 and D_2 are the diffusion coefficients of the reactant species and r_{12} is the reaction radius. The diffusion coefficient in water of growth hormone of 21 000 molecular weight is approximately 7.23×10^{-7} cm^2 s^{-1} [194]. The diffusion coefficient of the cell or membrane is unknown but is obviously less than the smaller molecule of growth hormone. The reaction radius for most reactions is about 5×10^{-8} cm. Thus the maximum rate constant for the hormone–receptor complex formation should be on the order of 10^8 1 M^{-1} s^{-1}, although the rate of association of growth hormone has been estimated [164] to be 1×10^5 1 mol^{-1} s^{-1}. The rapidity of response at sub-physiological hormone concentrations on the basis of spectroscopic observations [2, 263, 288, 310], would suggest a greater rate of association. From studies [279] of the *lac* repressor operator system, rates of association greater than those expected on the basis of free diffusion rate limits can be applied to hormone binding. Rather than an uncharged species, at physiological pH the growth hormone molecule has a net negative charge although the net charge of the receptor is unknown, but possibly positive. This could result in electrostatic interaction with accelerated association. In addition, the growth hormone may not simply diffuse randomly as it would in solution but rather there may be oriented diffusion. Rather than a three-dimensional random walk, the growth hormone may bind to the cell surface and may 'roll' or 'hop' along it, thus reducing the search of the hormone to only two dimensions. Moreover, the rate of association of the first few molecules of growth hormone may be significantly faster than subsequent molecules but could not be detected experimentally.

Growth hormone is not a unique ligand in causing changes in state of membrane proteins. The tripeptide, thyrotropin-releasing hormone, binds to receptors in the plasma membranes of the anterior pituitary gland [11] and stimulates release of the pituitary hormone, thyrotropin. The intrinsic tryptophan fluorescence of the membrane of pituitary cells was quenched by thyrotropin-releasing hormone [152]. Similar quenching was not produced with a pituitary cell line which was not a target for thyrotropin-releasing hormone as opposed to the cell line where receptors had been demonstrated. In addition, inactive analogs of this tripeptide ligand were without effect on the responsive pituitary cells. There was a suggestion of a dose response relationship, although the maximal changes of about 2.5% made statistical evaluation of such changes difficult. As with the growth hormone studies cited above, a small number of thyrotropin-releasing hormone molecules (10^{-9} M) with 10^6 to 10^7 pituitary cells were able to initiate a change in state of a large number of pituitary cell membrane protein molecules, i.e. one molecule of thyrotropin-releasing hormone per 10 000 tryptophan residues [152]. Since pituitary cells were pre-incubated for 5–20 minutes before fluorescent measurements were made it is not possible to state whether effects were demonstrable at times earlier than 5 minutes.

The studies discussed suggest that there are changes in secondary and tertiary structure of proteins when peptide hormones interact with the membrane. Moreover, small changes in state of many membrane proteins are more likely rather than major changes bordering on gross denaturation of a few proteins. From transmission, scanning and freeze-etch electron microscopy there are membrane changes on interaction of peptide hormones with target cell membranes which are indicative of possible changes in quaternary structure of proteins or aggregation of different proteins. The pituitary hormone, prolactin, in addition to its effects on mammary tissue and corpora lutea of the ovary, has been considered to have a role in salt transport through an effect on the kidney. By freeze-etch electron microscopy, prolactin was found [342] to increase the number of particles in the membranes of teleost kidney cells and there was a positive correlation between particle numbers and (Na^+/K^+) ATPase activity. Since the hormone was administered to the fish 10 hours before sacrifice, it is difficult to attribute this effect to a direct effect of prolactin on the plasma membrane. In view of the long time interval from hormone administration to membrane examination, the membrane changes may be a reflection of metabolic changes intracellularly in the kidney cell or in other tissues.

The posterior pituitary hormone, vasopressin, is known to have antidiuretic properties in mammals and non-mammals. In the urinary bladder of the toad, vasopressin will stimulate water reabsorption in addition to producing a number of membrane changes. When isolated toad urinary bladder, which is a known target tissue of vasopressin, was exposed to vasopressin for 30 minutes there resulted a stimulated osmotic water flow associated with distinctive and organized sites of aggregated intramembranous particles [163]. These particles were only found in granular cell luminal membranes and not in membranes of adjacent mitochondria-rich or goblet cells. There was clearly a linear relationship between the frequency of aggregation sites and vasopressin-stimulated osmotic water flow [163]. Exposure of bladder to vasopressin for less than 30 minutes was not reported. The aforementioned spectroscopic changes in liver [2, 263, 288, 310] and pituitary cell membranes [152] suggested changes in secondary and tertiary structure of many proteins in a few minutes or less. Intramembranous particles with diameters of 60–120 Å probably represent proteins or lipoprotein units in a hydrophobic medium [35, 71]. That such particles should aggregate or coalesce and that most of the membrane particles are involved indicates that there is a profound change in membrane structure in response to peptide hormone.

Another indication of widespread changes induced by hormone in the membrane comes from studies [93] which demonstrated that insulin at 10^{-8} for 1–2 hours produced profound changes of the surface membrane of serum-starved fibroblasts. Insulin produced an increase in the number of microvilli on the surface. This would indicate that multiple components of the cell membrane are involved and not merely membrane proteins. Since there were changes in intracellular components as well as the surface membrane and since the effect of insulin on isolated membranes was not studied it is not possible to distinguish between a direct effect of insulin on membrane components and the consequence of intracellular alterations, e.g. enhanced protein synthesis or tubulin aggregation. The appearance of microvilli on the surface membrane within one hour of exposure to insulin coincides with increased uptake of leucine, glucose, uridine [137, 177], and decreased cAMP [177, 299]. It may be that the decreased cAMP generated in the membrane as a result of insulin inhibition of adenylate cyclase or stimulation of cAMP phosphodiesterase activity has profound consequences in terms of membrane protein phosphorylation by cAMP-dependent protein kinase. Several protein hormones do influence plasma membrane protein kinase activity [17, 73, 101, 170, 207, 282, 296] some as rapidly as 1–2 minutes [282, 296]. This stimulation of membrane protein phosphorylation may have cAMP-dependent and -independent components. The regulation of plasma membrane protein kinase activity by vasopressin [73, 296] may affect only one specific protein which is related to the permeability changes produced by vasopressin [73]. A decrease in phosphorylation of this specific protein can be partially accounted for by this activation of a cAMP-dependent membrane-bound phosphoprotein phosphatase [73].

1.3.2 Membrane effects monitored by probes

Membrane perturbations produced by probes

In the above studies [2, 152, 263, 288, 310] of hormone–membrane interaction, changes in intrisinc properties have been employed to detect changes in state of proteins in target cell membranes. Additional information has been forthcoming from studies utilizing fluorescent and spin-labeled probes. *A priori* one would anticipate qualified interpretations of data derived from experiments which involve the introduction into membranes of components not normally present. This expectation has been borne out in the case of the growth hormone–liver membrane system where it has been found [263] that while the quantum yield of fluorescent 7, 12 dimethylbenzanthracene (1.6×10^{-6} M) in the liver membrane preparation has been changed by exposure to growth hormone, there was no alteration in the circular dichroism pattern as was noted [2, 263, 288] in the absence of the fluorescent probe. Lipoidal nitroxide spin probes have been found [26] to modify osmotic fragility and surface topology of erythrocytes. In the concentration range of $10^{-10}-10^{-5}$ M studied [26] the spin probes produced an initial osmotic stabilization and echinocytic morphology of the erythrocytes (10^{-6} M) followed at slightly higher concentrations ($10^{-5}-10^{-4}$ M) by osmotic fragility, sphering and erythrocyte lysis. Surface topology could be modified with as few as 2×10^4 molecules per cell. These observations [26] suggest that there are local concentrations of the spin probes which expand into the protrusions noted by scanning electron microscopy. Cell lysis may be the consequence of irreversible breakdown of membrane structure. The spin probes may not sample the whole area of a labeled cell. Although at low concentrations the spin probe is likely to be confined to the lipid phase, at higher concentrations of 6×10^{-7} M or 3×10^8 molecules per cell, there is appreciable binding to membrane proteins [26]. With human erythrocytes over a concentration range in terms of spin label/lipid ratio of 1:83 to 12:1 [43] compared to the range in the previous study [26] of 1:25 000 to 39:1, it was noted [43] that morphological changes were somewhat apparent in the 1:83 region with slight hemolysis. At a label/lipid ratio of 1:17 most cells showed morphological changes with slight hemolysis [43]. Although at a label/lipid ration of 1:10 or less the basic ESR spectrum was considered to be unaffected, there was an indication of a 4% decrease in the order parameter in going from a ratio of 1:83 to 1:8.3 suggestive of increasing fluidity of the membrane [43]. This should be contrasted with changes in the order parameter of 1% which have been noted with lipoidal nitroxide spin probes in erythrocyte membranes perturbed with hormonal ligands [150, 179, 180].

In erythrocytes, a spin-labeled fatty acid has been employed at concentrations of $10^{-5} - 10^{-4}$ M [150, 179] while in lymphocytes, concentrations of 3×10^{-4} M have been employed [13]. Further caution is dictated by the observations that the use of spin label and fluorescent label of the same parent compound, stearic acid, in pure and mixed monomolecular films lead to different estimates of fluidity [44]. In mixed films under membrane-like conditions, the spin label probe, 12-nitroxide stearic acid exhibited positive deviations and the fluorescent probe 12-(9-anthroyl) stearic acid exhibited negative deviations from ideality [44]. This would give readings of too high and too low fluidity respectively [44].

Not only structure but function may be modified by membrane-associated probes. The fluorescent probe, ANS, inhibited chloride and sulfate transport in human erythrocytes [107] at concentrations less than 1 mM, similar to those which produce crenation in these cells [106, 354]. This probe which appears to be widely distributed in the phospholipid and protein regions [132, 188] of the membrane at concentrations below 0.5 mM also partially inhibited the ouabain-sensitive sodium and potassium influx and efflux. In addition, ANS at concentrations greater than 1×10^{-4} M has been found [178] to increase the fluidity of dipalmitoyl lecithin vesicles as measured by pyrene excimer fluorescence. The sodium-potassium ATPase and ouabain-sensitive sodium and potassium fluxes are also inhibited by ANS. It has been noted [252] that greater than 5×10^9 molecules per cell of a nitroxide derivative of the sterol, androstane, commonly used as a membrane spin probe, will inhibit cholesterol synthesis in normal and leukemic cells. At concentrations below 5×10^9 molecules per cell, this spin label probe stimulates cholesterol synthesis [252].

Interpretation of such data will have to consider the perturbation, inhomogeneity of the probe, difficulty in associating the probe with a particular membrane component and consequences of probe–probe interaction. Without associated studies of other intrinsic properties of structure (intrinsic fluorescence, circular dichroism) or function, (enzyme activity, transport) which may be more sensitive to local perturbations, one would have to employ such probes with caution. With the above reservations, about spin label and fluorescent probes in mind, interesting results have been forthcoming about changes in membranes on interaction of the putative receptor and its ligand.

Hormones

The peptide hormone, angiotensin, has been shown to bind to plasma membranes from smooth muscle of guinea pig ileum [295], as well as to plasma membranes of smooth muscle of aorta [80]. When such membranes [295] are labeled with the spin probes, MSL-6 (4-maleimido-2, 2, 6, 6-tetramethylpiperidinooxyl), a 'protein-specific label', two components were apparently labeled on the basis of electron spin resonance analysis. One probe labeled sites of higher mobility and another was indicative of sites with greatly restricted motion [295]. Angiotensin (1×10^{-3} M) produced an increase in mobility of the labeled components in the more hindered environment [295]. Since the hormone was present in large excess compared to membrane protein, it is difficult to consider a concerted mechanism operating in these experiments. When the membranes were labeled with a 'lipid spin probe', 5-SASL-2-(3-carboxypropyl)-4, 4 dimethyl-2-tridecyl-3-oxozolidinyloxyl, ESR spectra of the stearic acid probe in the membrane suggested two different states. One had a high degree of mobility and in another, motion was hindered. Angiotensin produced a decrease in hyperfine splittings and line widths which were considered to originate in probes tumbling more freely in aqueous environments. In contrast to these changes in membranes similar changes were not noted [295] when angiotensin was added to an aqueous solution of the probe.

Human erythrocytes, when exposed to physiological concentrations of prostaglandins, epinephrine and carbamylcholine can have their flexibility altered [6, 150, 179]. This is consistent with the morphological changes noted [7] with low concentrations of prostaglandins PGE_1 or PGE_2, which can increase or decrease the internal volume of the cell. With only 1–10 molecules of prostaglandin per cell at concentrations of 10^{-12} – 10^{-10} M, the paramagnetic resonance spectra of spin-labeled fatty acid bound non-specifically to erythrocyte membranes showed changes suggestive of alterations in membrane structure [180]. Prostaglandins (10^{-12} – 10^{-5} M) also altered the circular dichroism spectra of erythrocyte ghosts without a dose response curve [224]. When erythrocytes were exposed to prostaglandin PGE_2 and epinephrine, decreases in fatty acid chain flexibility and erythrocyte bilayer fluidity were also correlated [180]. Carbamylcholine produced an increase in membrane rigidity as indicated by a 0.7–1.6% change in order parameter which was maximal at 10^{-6} M carbamylcholine [149]. The muscarinic blocking agent, atropine, antagonized this effect of carbamylcholine. The neurotoxin, tetrodotoxin, in the presence of a calcium chelating agent, also blocked this effect while cytochalasin B, a fungal metabolite which disrupts microfilament structure and inhibits processes mediated by contractile proteins produced a decrease in membrane rigidity [149]. The response to carbamylcholine required the presence of internal ATP and calcium. The changes noted [150, 179] with carbamylcholine may or may not be related to the co-operative structural rearrangement of bovine erythrocyte membranes produced by acetylcholine and studied by the circular dichroism method [338]. In the latter study [338] the effect was eliminated by inhibition of acetylcholinesterase activity implicating acetylcholine hydrolysis in the membrane perturbation. Moreover, the effects were demonstrable at concentrations of $10^{-3}-10^{-2}$ M, several orders of magnitude greater than the concentrations employed [150, 179] with carbamylcholine, a non-hydrolysable analog of acetylcholine.

These effects of the prostaglandins, epinephrine and carbamylcholine on the intact human erythrocyte and erythrocyte ghost raise questions about the presumed receptors and the biological effects of these ligands. Although specific interaction of norepinephrine with turkey erythrocyte ghosts has been reported [27, 28, 294], no stereospecificity was demonstrable and the binding appears to be selective only for the catechol function [27, 219]. Catecholamine receptors are now being found in chicken [206] and human erythrocytes, resealed erythrocyte ghosts and leaky erythrocyte ghosts [331]. Since major questions have been raised [61, 63] about β-adrenergic receptors in general, one will have to await more definitive data in this regard. Preliminary experiments [149] lend support to the presence of an acetylcholine receptor in human erythrocytes. They suggest that this 'receptor' is a protein of 41 000 daltons in contrast to the 91 000 dalton acetylcholinesterase. It would appear that specific receptors in the human erythrocyte membrane have not been described for prostaglandins and studies to date may merely reflect a general solubility in the lipid phase of the membrane. It may well be that there are a limited number of such receptors on human erythrocyte membranes as with catecholamine receptors in the chicken erythrocyte [206] and these are adequate to interact with carbamylcholine [150, 179] and prostaglandin [179, 180] with associated changes in membrane structure.

The changes in membrane structure noted with the catecholamines [150, 179] and a spin-labeled fatty acid may have its pharmacological counterpart. Propanolol, a drug with β-adrenergic blocking properties has been reported [89, 120, 211] to increase the permeability of human erythrocytes to potassium with very little change in sodium permeability.

Epinephrine, which in the above experiments [150, 179] changes membrane structure, has itself not been similarly studied.

To explain many of the observations of membrane changes of state associated with specific ligand receptor interaction one might consider changes in membrane fluidity. In one study [213], the influence of insulin on erythrocyte membranes of rats fed a corn oil- or lard-supplemented diet was determined. Presumably, the erythrocyte membranes exhibited high and low fatty acid fluidity respectively. Hill coefficients, *n*, for the inhibition by fluoride of acetylcholinesterase and (Na^+/K^+)-ATPase were obtained. The values for *n* for acetylcholinesterase were 1.5 for corn oil-fed rats and 1.0 for lard-fed rats. In the presence of insulin, *n* decreased from 1.5 to 1.0. Half-maximal effects of insulin were noted at 4.5×10^{-10} M within the normal range of plasma concentrations. No effect of insulin was noted on acetylcholinesterase of lard-fed rats where co-operative transitions were already minimal. Whereas solubilized membrane preparations had Hill coefficients similar to the native membrane preparations, there was no effect of insulin. Insulin did decrease the Hill coefficient with membrane-like reconstituted material. This would suggest a membrane-mediated effect on the co-operative behavior of acetylcholinesterase rather than a direct effect on the enzyme. Inverse effects were noted with (Na^+/K^+)-ATPase where insulin increased the Hill coefficient of corn oil-fed rats from 2.39 to 2.95.

These studies [213] suggest that insulin does alter membrane fluidity and as a consequence influences several enzymes in the membrane. Although no receptor for rat erythrocytes has been demonstrated, insulin has been shown to bind to turkey erythrocytes [118]. Presumably, after binding of insulin to the putative receptor of the rat erythrocyte there is a local change of state with subsequent propagation throughout the membrane. With a change in membrane fluidity produced by insulin many more proteins may have an altered change of state in addition to the two enzymes studied [213]. This would be consistent with the conclusions drawn from spectroscopic studies [152, 224, 310] with other hormones.

Surface modifications of the epithelial cells of amphibian urinary bladder have been produced by anti-diuretic hormone for which there is a specific receptor in this tissue. This has been demonstrated [255] by concanavalin A-mediated hemadsorption to epithelial cells at 22°C, the mean body temperature in these amphibians. The hormone did not influence the concanavalin A-mediated hemagglutination at 4° or 37°C.

At the latter temperature there was already enhanced hemadsorption to the untreated cells with elevation of temperature. Since concanavalin A-mediated agglutination may require lateral mobility of lectin receptors in the plane of the membrane [236], the augmented agglutination at 22°C of cells treated with anti-diuretic hormone would suggest an effect of the hormone on the fluidity of the membrane. In these experiments [255] the analytic procedure utilized a lectin probe, a cell surface labeling agent which can cause rearrangements of randomly distributed membrane components [41]. Since removal of the label bound to the cell surface allows the membrane components to return to their original distribution [41], it would be important to know that the hormone effect persisted after removal of the label. This, of course, would require monitoring of an intrinsic property or another less perturbing probe.

Acetylcholine

For complete understanding of how the message of specific ligand–membrane interaction is transduced into some change in structure of the membrane, it will be necessary to have some definition of the relationship of the specific receptor to other functional units related to the ligand. Some progress has been made in this regard with acetylcholine, the specific binding of which has been demonstrated in the electric organs of the teleost *Electrophorus* and elasmobranch, *Torpedo* (summarized in [50,166]).The electroplax of both *Electrophorus* and *Torpedo* are asymmetrical cells which are innervated on but one face. It has been estimated that this innervation involves 1–2% of the plasma membrane surface in the case of *Electrophorus* [31] and 50% in the case of *Torpedo* [153, 300]. From a functional point of view it may be that there are one or two acetylcholine-activated receptor sites per available ion channel [4]. In addition, acetylcholine receptor-rich membrane fragments with a binding site for inorganic cations [226] may be reassociated with membrane lipids and a chemically excitable membrane is formed [133, 225, 226]. Thus the acetylcholine binding subsite contains the elements necessary for ion translocation. Whether a single membrane protein is responsible for both ACh and ion binding is not clear at present.

The spin-labeled long chain acylcholine, 8-doxyl-palmitoylcholine, interacts with the cholinergic receptor protein of *Electrophorus* membrane fragment and shows an electron spin resonance spectrum indicative of complete immobilization [39]. When membrane fragments were incubated with stoichiometric amounts of cholinergic ligands such as carbamylcholine, decamethonium or *d*-tubocurarine, the mobility of the label was enhanced suggesting that the cholinergic ligands interfered with the specific interaction with receptor protein and redistribution into the more fluid lipid phase of the membrane [39]. Without an independent measure of the lipid phase of the membrane it is difficult to know whether the cholinergic ligand produced a change in the fluidity of this region. It was early reported [167] that the binding of the commonly used fluorescent probe, ANS (1-aniline-8-naphthalene sulfonate) to excitable membranes of eel electroplax was increased by *d*-tubocurarine or flaxedil, two receptor inhibitors, while there was no effect of carbamylcholine or decamethonium, two receptor activators [167]. Although the affinity changed there was no change in the number of binding sites [167]. Although this would indicate a change in state of membrane proteins, it does not suggest which proteins or the nature of the protein changes. When the binding of the fluorescent cholinergic analog DNS-chol (1-(5-dimethylaminonapthalene-1-sulfonamido) propane-3-trimethylammonium iodide) to electroplax membrane was studied it was noted [49] that the blue spectral shift characterisitc of the presence of agonist was lost when an antagonist was substituted for the agonist. It was concluded [49] that these spectroscopic changes were indicative of a change in membrane structure which is associated with the physiological response of the membrane. Since DNS-chol interacted with 'secondary' (lower affinity) membrane sites as well as the cholinergic receptor site and there was a lower efficiency of energy transfer from membrane protein to DNS-chol, it may be that these 'secondary' sites lie close to but are not identical with the cholinergic receptor site. In view of the fact that with the addition of carbamylcholine there was a change in extrinsic fluorescence but not intrinsic protein fluorescence ($\leqslant 2\%$) of membrane fragments [49] it would appear that the fraction of protein components the conformation or environment of which changed was perhaps limited to the receptor protein.

With purified detergent solutions of the receptor protein of *Electrophorus* there was a 5–10% decrease in intrinsic fluorescence with cholinergic agonists, while antagonists produced either no change or an increase in intrinsic fluorescence [49]. However, with the receptor protein of *Torpedo*, no change in intrinsic fluorescence was noted with the cholinergic ligands at 10^{-7}–10^{-5} M in the 1–5 s or ms range [90].

Colicin

There are a number of *E. coli* membrane responses associated with specific ligand interaction which suggest that these are similar to the membrane responses to hormones.

The colicins are bacteriocidal proteins produced by certain strains of Enterobacteriaceae and are active against other strains. Colicin E_2-P9 has been purified to chemical and physico-chemical homogeneity [204]. It is a simple protein with a molecular weight of about 60 000 [136, 240]. Colicinogenic cells are resistant to the killing action of the colicins they produce. Although receptor specificity was initially determined by isolation of resistent bacterial mutants, actual binding of colicin to the surface of sensitive cells but not resistant cells has been demonstrated [204] using purified radioactive colicins at saturating levels. It has been demonstrated [182] that colicin E_1 binds to the cell surface, and with colicins E_2 and E_3 all or part of the molecule must penetrate the cell for the killing action to be observed.

One sensitive cell bound 11–90 killing units of colicin [204, 217, 273]. A killing unit is defined as the amount of colicin necessary to kill a single sensitive cell as measured by the number of colony-forming survivors. Since one killing unit corresponded to about 1 [273] to 100 [204] molecules, the number of receptor sites varied from 11–9000 [240]. The killing action of most colicins is on the average a single-hit process [154, 217, 273] as determined from experiments where cells were treated with less than saturating amounts of colicin and the number of colony formers were measured. A plot of the logarithm of the fraction of survivors against the amount of colicin added was consistent with the single-hit hypothesis. The number of killing units may or may not correspond to the actual number of adsorbed molecules [204, 273], but as noted above with amounts of colicin which are much less than needed to saturate all receptor sites, the actual number of bound molecules is likely to be less than 100 and may be as few as 1. Thus, as previously noted for erythrocytes, erythrocyte membranes and liver membranes with growth hormone [288, 310] and prostaglandins [150], a small number of specific ligand molecules bound to the receptor is capable of initiating profound changes, in the case of colicin leading to death of the sensitive cell.

Although the binding of 1 to 100 molecules causes some observable biochemical effect, such changes appear slowly, i.e., in minutes. The observable changes may appear more quickly when the dosage of multiplicity or number of killing units of colicin is increased. The multiplicity is calculated from the number of survivors after the adsorption of colicin is complete, i.e. 15 minutes. Although the binding may be complete within 15 minutes, one would like to observe changes in the membrane before this time since there would be significant interaction of colicin with the bacterium at earlier times and according to the single-hit hypothesis [240] each interaction results in the killing of a sensitive cell.

When *Escherichia coli* were labeled with the fluorescent probe, ANS, the addition of colicin E_1 to sensitive cells resulted in a rate of fluorescence increase which was a linear function of the multiplicity of colicin [57] and an 8 nm blue shift [253]. On the basis of cell survival the colicin binding was more than 99% complete within one minute. Although there was a lag of about 30 seconds in the fluorescence response, more than 50% of the increase occurred within one minute [57]. It has been possible to observe from the kinetics of colicin binding as determined by cell survival and the rate of fluorescence increase, that binding precedes the fluorescent probe response and the associated structural changes in the cell surface [253]. The further fluorescence increase occurring after one minute may have been related to a process which followed binding. The transition for zero response to maximal response came at a colicin E_1 multiplicity of less than two with a ratio of colicin molecules to *E. coli* of no more than 100 to 1 [57]. The maximal change of fluorescence occurs with a single killing unit. Thus the change of fluorescence shows an 'all-or-none' response and is consistent with the single-hit mechanism. Interestingly there was a similar rate of ATP decay and fluorescence increase suggesting close relationship of the two processes. However, it is possible to note an increase in fluorescence of sensitive cells with colicin in the absence of any decrease in intracellular ATP or any large decrease in potassium level [253]. This would suggest that the membrane conformational change is not the consequence of the metabolic changes.

Whereas ANS is considered to probe the region of the membrane between the lipid bilayer and the aqueous phase, the fluorescent N-phenyl-1-naphthylamine (NPN) probes the more lipophilic region of the membrane. In sensitive *E. coli*, colicin induced a fluorescence increase with NPN similar to that noted with ANS [253]. This would suggest that more of the cell membrane than only the region sampled by ANS is involved in the conformational changes produced by colicin. The lag period and the initial linear rate of fluorescence increase were temperature-dependent [58]. Arrhenius plots of the rate of fluorescence increase were biphasic, the transition temperature being dependent on the culture growth temperature and consequently the fatty acid composition. There is an increase in saturated fatty acids at the higher culture temperatures with consequent increase in melting point . From fluorescence polarization experiments with NPN [58], it would appear that the transition temperatures of the Arrhenius plots are associated with a disordering of the cell membrane. The colicin E_1 effects appear to be associated with a disordering of fatty acid chains in part of the cell membrane.

It is reasonable to ask what processes would be initiated by colicin E_1 to produce a disordering of the hydrocarbon fatty acid chains as reflected in a decrease in activation energy. Perhaps colicin E_1 moves laterally through the hydrophobic core by diffusion, as has been suggested for cholera toxin in eukaryotic membranes [20]. It should be noted, however, that lateral diffusion constants for membrane proteins are $\simeq 10^{-9} - 10^{-12}$ cm^2 s^{-1} and a membrane response within seconds would be inconsistent with lateral diffusion requiring minutes to hours (see Section 1.2.1). Perhaps we should look for some chemical modification of lipids which might be triggered by an activation of phospholipase near the site of colicin binding. However, an enzymatic reaction might be slower still. Under conditions where colicin E_1 leads to increases in fluorescence intensity with a maximum half value in 2 minutes, there was little change in lysophosphatidyl ethanolamine levels for 30 minutes [56]. Evidence in this system could not be adduced to involve phospholipase A or D in the early biochemical effects of colicin E_1 [56].

Without evidence of specific enzyme activation e.g. phospholipase A or D, attention has been directed to colicin-produced membrane events associated with divalent cation changes. When colicin K was added to a system containing sensitive *E. coli* cells and chlorotetracycline, in the absence of calcium, there was a two-fold enhancement of fluorescence and a three-fold decrease in fluorescence anisotropy [37]. The chlorotetracycline fluorescence indicated the involvment of magnesium. The fluorescence began to change after a lag of about 20 seconds and could be demonstrated with a colicin multiplicity of one [37]. As with studies of ANS fluorescence [239], the colicin effect was of an 'all or none' response of a bacterial cell. Although the stimulated efflux of 1-D-galactoside by colicin at a multiplicity of 4 paralleled the rise in fluorescence, the former was more rapid being half maximal at 1.3 minutes compared to a half maximal time for fluorescence of 4.2 minutes. This would suggest that the chlorotetracycline fluorescence is secondary to some other membrane events. Indeed, with the use of radioactive tetracycline it was established that colicin K caused an increased uptake of the probe rather than a change in quantum yield.

Perhaps a more proximal membrane action of colicin may be related to its interference with an energy-coupling mechanism for certain transport systems [102, 203, 259]. No effect on chlorotetracycline fluorescence response to colicin was noted [37] when N, N′-dicyclohexylcarbodiimide, an ATPase inhibitor, was present. This agent blocks the decrease in cellular ATP by colicin without affecting transport [98]. This is similar to the colicin-produced changes in fluorescence of N-phenyl-1-naphthylamine in the presence of the same ATPase inhibitor which prevented ATP changes [253]. Since carbonylcyanide *p*-trifluoromethoxyphenylhydrazone and 2, 4-dinitrophenol, uncouplers that presumably act by abolishing the membrane proton gradient [144], block colicin-stimulated fluorescence of chlorotetracycline, it would appear that electron transport is necessary to form a membrane-energized state and for colicin to act on the membrane or link its action to other membrane processes. However, the high intracellular concentration of magnesium would suggest that the accumulated probe is in the cytoplasm or on the inner surface of the membrane. Perhaps the membrane of colicin-treated cells becomes more permeable to magnesium and the energization and de-energization of the membrane may affect magnesium partitioning. In view of the decrease in fluorescence anisotropy with colicin [37], of interest is the observation [328] that magnesium addition to phospholipid bilayers induced a liquid to liquid crystalline phase transition of fatty acyl chains. This raises the broader question of the possible role of transmembrane potential or pH gradient in the control of membrane fluidity and the effect of specific ligands for the cell surface. In this regard is the oft-noted colicin inhibition of potassium transport [343] where a single lethal hit essentially empties a cell of potassium. Although the half time of maximal loss of potassium at a multiplicity of 300 is about 0.5 minutes and about 1.5 minutes at a multiplicity of 4 [343], it is difficult to compare these experiments with those [37] done under somewhat different conditions.

More direct information of the role of transmembrane potential comes from studies [38] with the potential sensitive fluorescent probe 3, 3-dihexyloxacarbocyanine. The bacterial protein, colicin K, produced a doubling in fluorescence of this probe in sensitive *E. coli*. This is consistent with a depolarization of the transmembrane electrical potential. This was associated with a rapid efflux of radioactive rubidium, a potassium analogue. The probe fluorescence could be further increased by anoxia or cyanide and decreased by glucose and oxygen. Thus the specific ligand, colicin K, does not completely depolarize the membrane potential. That component of membrane energization associated with electron transport was unaffected by colicin K. Of significance in this study [38] was the fact that the observed effects were noted at multiplicities of colicin K of 2 to 4. Fluorescence changes were noted within 15 seconds in the presence of glucose and 30 seconds in the absence of glucose. Glucose increased the potential of energy-deprived cells. Anoxia and cyanide, respiratory inhibitors, reduced the membrane potential significantly. With amounts of colicin K above two killing units per cell, a maximum probe fluorescence was reached with additional colicin K causing no further increase in fluorescence. This is consistent with the irreversible 'all-or-none' effect.

Membrane depolarization by colicin could result from a transmembrane flux of cations and/or anions since the fluorescence rise could occur in the absence of extracellular permeable cations [38]. In addition, cells treated with colicin K lose their internal potassium [343] and organic anions [103]. The kinetics of cation [$^{86}Rb^+$] efflux was similar to the rise in probe fluorescence [38]. The rise in fluorescence was not due to Rb^+-K^+ influx. In fact, the rise in fluorescence by colicin K is inhibited by extracellular KCl over the range of 4–50 mM [38]. Similar effects were noted [38] with sodium. Although phosphate can be actively transported to an internal concentration of 10 mM [221], colicin K only increased phosphate transport slightly [38], suggesting that this transport is not dependent on membrane potential. Proton permeability [98] and intracellular pH [38] were not significantly altered by colicin in sensitive *E. coli*. Thus, membrane depolarization by colicin K was unrelated to proton gradient alteration.

It is important to establish whether the 'membrane potential' probe, carbocyanine, is indeed sampling a change in membrane potential in response to colicin or is merely reflecting a new probe environment independent of membrane potential. Other fluorescent membrane probes [37, 58, 134, 253] as noted, respond in a similar manner to colicin. Although the membrane potential may not be a factor in the fluorescence of one of these probes, N-phenyl-1-naphthylamine [58, 134, 253], colicin could affect the fluorescence as a consequence of a change in state of membrane lipids associated with an overall change in membrane architecture.

1.4 ION TRANSLOCATION

Cell surface active hormones are similar to neurotransmitters in several respects. Like acetylcholine, some cell surface active hormones e.g.

insulin, growth hormone do not stimulate cAMP formation and may operate through other mechanisms involving ion permeability changes, cGMP formation and/or cAMP destruction. Most prominent for both hormones and neurotransmitters are the stimulated ion flows and changes in membrane potential soon after ligand binding to its receptor.

Hormonal effects on permeability to ions which determines membrane potential have been studied in the liver, a target tissue for two cell surface active hormones, glucagon and epinephrine. For this purpose, perfusion of the intact liver with added hormones and incubation of liver slices with added hormones and cyclic nucleotides have been employed. In the isolated perfused rat liver, glucagon or cyclic AMP administration was followed by a biphasic response in the flow of monovalent cation i.e. potassium, sodium and hydrogen [108]. There was an initial liver uptake of potassium and sodium with slight perfusate pH decrease at 2.5 minutes followed by a subsequent larger loss of sodium and potassium from the liver, maximal at 6–7 minutes. The perfusate pH increased at these later times. Pefusate radioactive calcium did not show a biphasic response and reached a maximum at about 4 minutes and was followed by increased glucose production [111]. Hyperpolarization of the liver membrane measured with microelectrodes occured at approximately the time of maximal potassium and sodium efflux i.e. 6 minutes [110]. Interestingly, omission of sodium, but not potassium or calcium from the perfusate prevented the glucagon- or cyclic AMP-induced glycogenolysis and gluconeogenesis but not the glucagon-induced increase in cyclic AMP [108, 111]. In the absence of perfusate sodium, the membrane potential was lowered and the cyclic AMP-induced calcium efflux was not blocked [64]. Similarly, isoproterenol produced hyperpolarization in perfused rat liver [112]. It should be noted that some substrates are also able to produce hyperpolarization of the liver membrane although not as promptly as cyclic AMP and of a more sustained and different ouabain sensitivity [65]. Substrate-induced membrane potential changes appear to involve mechanisms different than those of cyclic AMP. Since many of the agents employed in these studies have profound hemodynamic effects [241, 304] in the liver perfusion system, it is difficult to evaluate such studies, particularly with regard to kinetics [129, 241].

The hormonal effects on the hemodynamics of the perfused liver can be obviated with the use of liver slices and hormones added to the incubation medium. In the superficial cells of isolated mouse liver segments, small concentrations of glucagon, epinerphrine and cyclic AMP caused a dose-dependent membrane hyperpolarization which was apparent within

30 seconds after exposure to hormones and cyclic AMP [249]. The glucagon effect on the cyclic AMP level was also apparent within about half a minute, but not sooner, after start of glucagon perfusion [94]. Although these findings are consistent with the glucagon effect being mediated by cyclic AMP, an independent stimulation of membrane permeability is not excluded. With guinea pig liver slices, norepinephrine (10^{-6} M) produced a reversible hyperpolarization in less than 1 minute [129] and increased not only the efflux of $^{42}K^+$ but within 2 minutes caused the tissue potassium to fall 8% and the sodium content to rise [130]. Although the changes in tissue content of sodium and potassium were largely transient, it is noteworthy that the ratio of potassium to sodium decreased significantly by 20% within one minute and returned to normal in 8 minutes [130].

Hyperpolarization has also been noted in other tissues in response to hormones. Thyrotropin and cyclic AMP produced hyperpolarization of isolated thyroid cells [15]. With intracellular micro-electrodes, epinephrine produced hyperpolarization of parotid acinar cells [250, 251]. Carbamylcholine and phenylephrine produced biphasic increase in potassium release in rat parotid gland slices and the early transient potassium release did not require external calcium [264]. This corresponded to a transient hyperpolarization. The release of potassium ions from salivary glands is mediated by α-adrenergic receptors [212, 264] and may involve the ouabain-sensitive (Na^+/K^+)-ATPase [212].

Although a change in membrane potential may be associated with hormone administration, depolarization by means other than hormones need not be associated with a biological response. This would suggest that the transport of specific ions regardless of effects on membrane potential or other effects of the hormone are important biological determinants. In perfused rabbit adrenal glands, adrenocorticotropin (ACTH) altered membrane potential in a manner dependent on the ionic medium [215]. In a potassium-free medium, ACTH produced action-potential-like changes in membrane potential [215]. These had a mean amplitude of 46 mV and spike duration of 600–1800 ms. The steroidogenic responses to ACTH could be inhibited without effect upon membrane potentials. This had been similarly noted with acetylcholine-stimulated release of amylase by pancreatic exocrine cells [214].

The influence of hormones on ion transfer and membrane potential suggests a possible effect on (Na^+/K^+)-ATPase. It has been reported that epinephrine and glucagon reduced *in vitro* the activity of (Na^+/K^+)-ATPase of plasma membranes isolated from rat liver and the epinephrine effect was blocked by propranolol [202]. Although insulin alone did not modify (Na^+/K^+)-ATPase activity, it did counteract the effects of epinephrine and glucagon [202]. Because cyclic AMP had an effect similar to glucagon and epinephrine it was speculated that the effects were mediated by this nucleotide [202]. It has also been reported that insulin stimulates human lymphocyte plasma membrane [126] and rat uterus (Na^+/K^+)-ATPase [199]. There is the provocative observation of an association between (Na^+/K^+)-ATPase activity and the number of intramembraneous particles in teleost kidney cells induced by prolactin administration [342].

It is obviously important to establish not only the relationship between hormone receptor in the membrane and membrane transport protein but to specify the array of inorganic cations and anions as well as organic molecules whose transport is changed. Although changes in sodium, potassium, calcium and chloride flux have most often been associated with changes in membrane potential, other inorganic cations and anions e.g. magnesium, phosphate may be important as well for subsequent membrane and cytoplasmic events. Thus, Colicin K, in the absence of calcium produced an increase in magnesium uptake in 20 seconds at a multiplicity of one [37]. It is also possible that the increased transport of some organic molecules e.g. nucleotide precursors, is also important for altered membrane activities and may even precede inorganic ion changes as noted with Colicin on *d*-galactoside transfer [37].

In addition to the effect of hormones on monovalent cation transport, membrane potential and (Na^+/K^+)-ATPase activity, there is ample documentation of the translocation of divalent cations, particularly calcium [269] and to a lesser extent magnesium [199]. For example, the hypothalamic hormone, luteinising hormone-releasing hormone stimulated calcium efflux from rat anterior pituitaries associated with luteinising hormone release [344]. In rat aortae, noradenaline increased the rate of uptake of calcium into the calcium fraction resistant to displacement by lanthanum [121]. Calcium ions are transported across the cell membrane and have widespread effects intracellularly. In addition, the hormone may act on the membrane to alter membrane binding of these ions. Thus, exchangeable calcium binding was increased when physiological concentrations of insulin were incubated with adipocytes but not adipocyte membranes [218]. Stable pools of calcium and magnesium were identified in these membranes, but were unaffected by insulin treatment. The increased exchangeable calcium binding may have been associated with the increased phosphorylation of specific adipocyte plasma membrane proteins induced by insulin [47]. Pharmacological doses of insulin decreased and glucagon increased calcium binding to liver plasma membranes [303].

The aforementioned experiments concerned with the effect of cell surface active hormones on ion transport suggest a direct effect of the hormone on transport. With those hormones affecting adenylate cyclase, they do not exclude an indirect effect via cyclic AMP since the latter nucleotide can simulate many of the hormonal effects. This is emphasized by observations in the perfused rat liver which indicate that protein kinase is activated by cyclic AMP before there is a significant efflux of calcium, sodium and potassium [109]. Membrane phosphorylation occurred in a sodium-free environment or with tetracaine-inhibited ion fluxes.

That metal ions play a part in the control of enzyme and metabolic activities has long been accepted. Of particular concern to this discussion is the influence of these ions on transduction of the hormone message. We have already alluded to divalent cations as important determinants of membrane structure. With regard to membrane function, ions have effects on both enzyme activity and ion transport. Thus, with regard to the pH dependence of the adenylate cyclase reaction, it has been observed that in the absence of glucagon, adenylate cyclase increases with increasing pH up to a limiting value [195]. With glucagon the rate reaches a maximum and then decreases [195]. Although in the perfused liver system [108] cyclic AMP, like glucagon, produced changes in pH, it is difficult to establish enhanced cyclase activity as being due to pH changes. Magnesium, like pH, influences adenylate cyclase activity in the presence and absence of glucagon in a complex fashion [195, 275]. Calcium inhibition of (Na^+/K^+)-ATPase in membranes of kidney cortex and erythrocytes was due to competition with (Mg^{2+})-ATPase [92]. In addition, human erythrocyte membrane permeability to sodium and potassium has been reported to be under the control of internal calcium [283]. The role of sodium and potassium in membrane structure is less apparent, although alterations in ionic strength are known to affect membrane integrity [278]. These ions are obviously important for the membrane-bound enzyme (Na^+/K^+)-ATPase.

A change in the relationship of intracellular to extracellular ionic composition and associated change in membrane potential could have a profound effect on the state of several membrane components e.g. protein conformation, fatty acid flexibility. This change of state and/or the availability of an altered ionic environment may facilitate binding of specific ions e.g. calcium, magnesium to certain membrane-bound enzymes such as adenylate cyclase, ATPase, protein kinase, phosphodiesterase. The change in state of several membrane components and/or the availability of organic molecules as enzyme substrates or co-factors may facilitate interaction of these components resulting in enzyme activity. The changes in hormone receptors themselves may either be a reflection of an overall change in membrane organization as described above (Section 1.3) or may represent a change in the nature of the receptor itself as a result of membrane potential change. Depolarization of neurblastoma cells inhibit scorpion toxin binding by increasing the rate of dissociation and this is unrelated to the depolarization agent used [45]. Of interest in this regard are the observations that long-lived conformational changes in biopolymers may be induced by applied electric fields equivalent to those noted across many cell membranes [234]. In addition, with similar electric fields, membrane changes in chromaffin granules were observed [285] with relaxation times of several properties between 150 μs to 1 s.

The mechanism of the effect of altered ionic environment and membrane potential on the state of membrane components could essentially be an effect of specific ion interaction, e.g. calcium, magnesium, sodium, potassium, hydrogen. There is ample precedence for conformational changes in proteins due to pH changes and electrostatic interactions between charged groups on proteins and other molecules e.g. other proteins, phospholipids. This has been discussed in a general manner [345] and more specifically with regard to proteins [66] and lipids [1, 48].

1.5 THE MECHANISM OF HORMONE ACTION AT THE CELL SURFACE

The data reviewed above on receptor topography and hormone-induced spectroscopic changes and ion fluxes are clearly relevant to the mechanism of hormone action at the cell surface. Taken together with existing binding data and information on the hormone stimulation of several membrane-bound enzyme systems, these should provide the key pieces of information required for the formulation of a model of hormone action. We first briefly summarize the salient points reviewed in Sections 1.2–1.4 above and then outline a speculative model of the way membrane receptors might transduce the hormone message. We are quite aware that the data are still too fragmentary for a detailed molecular model of hormone action. Yet recent advances in the molecular dynamics of cell membranes and current data on the striking similarities between the membrane response to hormones and neurotransmitters allow for a fresh perspective on the mechanism of hormone action.

1.5.1 Biophysical contributions

Mobility and topography of hormone receptors

The last few years have seen a significant broadening in our thinking about cell surface structure and how receptors may fit into a dynamic membrane. The fluid mosaic model has been amended by recent data on microtubule and microfilament regulation of mobility of membrane components and by the state of membrane-bound water and cations and their implications for function. The kinetic constants for the various types of motion of membrane components have been delineated, and these will provide both a guide to and constraints on future molecular models of receptor function.

The various ultrastructural studies aimed at mapping hormone receptors have given a fairly consistent picture of the relatively few systems studied. Melanocyte-stimulating hormone, acetylcholine, and luteinising hormone have their receptors concentrated in large clusters or patches on the cell surface, while insulin receptors exist both singly dispersed and in small clusters of up to about 10 receptors. In contrast to cholera toxin receptors, there is currently no evidence for hormone-induced receptor aggregation or clustering. Indeed, as noted in Sections 1.2.1 and 1.2.2 hormonal activation of adenylate cyclase [94, 281] and alteration of membrane potential [129, 249] within seconds is inconsistent with lateral diffusion of membrane receptors in minutes to hours. In fact, the limited evidence to date would suggest that hormone receptors exist in a clustered or patched state independent of their occupancy by hormone [See Section 1.2.2].

These observations are relevant to various proposed models [19, 20, 72, 74, 75, 127, 156, 164, 190, 192, 219] aimed at interpreting the binding and enzyme activation properties of hormone receptors. Most models have attempted to account for a variable binding affinity and dissociation rate which are dependent on the amount of hormone bound, i.e. a non-linear Scatchard plot. Frequently low and high affinity states of the receptor have been postulated with interchangeability between receptor states. This dynamic equilibrium between different affinity and possibly different cyclase-activating states of the receptor can be shifted in either direction depending on the amount of hormone bound [72, 75, 127, 156, 192] and also by cations, pH, and guanyl nucleotides [127]. Some models have proposed negative co-operativity with 'site–site' interaction to account for the non-linear Scatchard plots [74, 75, 190, 192]. The molecular interpretations of this 'site–site' interaction have been described either in terms of an allosteric mechanism involving subunits of the receptor molecule (similar to hemoglobin-O_2 binding) [74, 75] or in terms of homone-induced clustering and interaction of receptor molecules with each other [74, 190]. An alternative set of models has been based on the lateral mobility of receptors in the plane of the fluid membrane [19, 20, 72, 156, 219]. These models generally view the hormone as binding to a free (unclustered) receptor and the hormone receptor complex subsequently interacting with an effector such as adenylate cyclase via lateral diffusion in the membrane. The hormone can also bind with a different affinity to a receptor–effector (e.g. cyclase) complex, thus providing at least two affinity states of the receptor.

None of the above models are completely consistent with the recent data on the topographical distribution of hormone receptors. The data for hormones (see Section 1.2) suggest that the hormone does not bind to the free (unaggregated) state of the receptor and then induce clustering. Rather, the hormone receptors appear to be in a clustered state prior to labelled hormone binding. One might argue that the basal level of endogenous unlabeled hormone might have induced the clustered state of receptors subsequently observed in ultrastructural studies. However, the recent observations [147, 148] of luteinising hormone receptors remaining clustered after hypophysectomy would not support this argument. On the other hand, reports of acetylcholine receptors spreading out of their concentrated patches below synaptic areas following denervation [128] might be consistent with a hormone-induced receptor aggregation. Further receptor mapping studies at very low endogenous hormone levels prior to incubation with labeled hormone should clarify this point. As far as models involving hormone-induced subunit interaction within a receptor in the membrane [74, 75], there is no physical or molecular evidence in support of these models, although a preliminary report indicates an insulin-induced dissociation of solubilized receptor subunits [164].

If we assume the receptors are clustered prior to hormone binding, then it is easy to imagine such a cluster as representing sites on a lattice or 'subunits' of a supramolecular receptor cluster. Events following binding (see Section 1.5.2 below) could influence neighboring bound hormone–receptor sites and alter their dissociation rates and binding affinities. Increasing receptor occupancy would increase the frequency of occupied site interactions and subsequently alter the binding affinities and dissociation rates of bound hormone. Such a mechanism might be consistent with an Ising lattice subunit model of membrane co-operativity [138]. Alternatively, alterations in the binding properties could result from the cluster or lattice as a whole being affected by hormone induced ion fluxes or a global change in state of membrane components as reflected by spectroscopic changes.

Spectroscopic changes in membrane

It is apparent that specific ligands produce widespread changes in the membrane including a large number of components. There is evidence that there are changes in state of many proteins as indicated by changes in circular dichroism [224, 310], intrinsic fluorescence [2, 152, 263, 288] electron spin resonance [150, 179, 180, 295] and electron microscopy [93, 163, 342] with a variety of cell surface active hormones (ligands) and neurotransmitters e.g. growth hormone, thyrotropin-releasing hormone, angiotensin, acetylcholine. This would be consistent with the changes in biological activity of several proteins e.g. receptor (summarized in [164]), transport (see Section 1.4), enzymes. In addition to adenylate cyclase, other membrane bound enzymes e.g. (Na^+/K^+) [202, 126, 199, 342] and Mg^{2+}-ATPase [2, 159], protein kinase [17, 73, 101, 170, 207, 282, 296], phosphodiesterase [200] and guanylate cyclase [151] are influenced by specific hormone interaction with the target cells. Membrane lipid components are also modified by cell surface active ligands reflected in changes of properties of hydrophobic fluorescent [2, 263] and spin-labeled probes [150, 179, 180, 295]. Moreover, the interaction of the lipid and protein regions considered to be sampled by a probe like ANS [132, 188] is also altered by the specific binding of ligand to membrane.

Not only are the changes widespread but there is evidence to suggest that changes in an interacting system are being reflected. These changes may actually be co-operative as indicated by the non-linear Scatchard binding plots (Summarized in [164]) and familiar sigmoid dose response curve of enzyme activation and possibly fluorescence [152]. Some of the physico-chemical changes are suggestive of an 'all or none' response in the sense that a relatively few molecules were able to elicit the maximal response. This was first suggested for hormone in the case of growth hormone on the basis of circular dichroism [2, 263, 288, 310] and intrinsic [263, 288, 310] and extrinsic [2, 263] fluorescence response to very small amounts of hormone. Changes, in intrinsic fluorescence of pituitary cell membranes with very small concentrations of thyrotropin-releasing hormone [152] were consistent with a co-operative effect. In addition, other cell surface active ligands i.e. catecholamines, carbamylcholine and prostaglandins in very small concentrations produced changes in the order parameter by electron spin resonance [150, 179, 180] or in deformability [6, 7] of the erythrocyte membrane. Essentially, the number of molecules in the membrane demonstrating changes in state was significantly greater than the number of bound ligand molecules. This is most dramatically illustrated by the effect of colicin on sensitive *E. coli* where it has been reported (summary in [240]) that a single hit of a toxin molecule with the cell causes death of the cell. Although there was a suggestive dose response relationship with thyrotropin-releasing hormone on sensitive pituitary cell membranes [152] and a biphasic response was reported with bovine growth hormone and liver cell membranes [288], there are insufficient data in this regard. It is important to note that the changes in state of the various membrane components occur very rapidly i.e. within seconds to minutes. Because of the experimental limitations of other methods this has been most clearly demonstrated by spectroscopic methods. These changes appear to occur as rapidly or more rapidly than changes in enzyme activation [94, 281]. Plasma membrane adenylate cyclase has been reported to respond to hormones in a period of seconds [94, 281].

Some of the spectroscopic changes in the membrane resulting from interaction of a ligand with its specific receptor are small e.g. 1–5%. These are generally average changes of several components e.g. secondary or tertiary structure of proteins, flexibility of fatty acid chains, etc. This would be consistent with changes in function of receptors, enzymes and transport proteins noted with hormones. Although fluorescent and spin label probes may sample specific compartments, generally, such probes do not reflect changes in a specific component. A possible exception might be a labeled ligand which binds to a specific receptor. Although, fluorescent substrates for enzymes have been studied, they have not been widely applied to study membrane phenomena [2].

The rapid and widespread appearance of changes in properties of several membrane components associated with hormone interaction with its target membrane naturally raises questions about possible mechanisms of such effects. Since, in excitable tissue exposed to neurotransmitters spectroscopic changes appear and spread over the entire membrane within seconds, it is tempting to suggest that these changes are related to changes in membrane potential as a reflection of changes in ion translocation or permeability.

Ion translocation
There is ample documentation that cell surface active hormones, including neurotransmitters, produce significant movement of ions including sodium, potassium, calcium and magnesium with associated changes in pH and membrane potential [357] (Section 1.4). Many of these changes in ion transport can similarly be produced by the cyclic nucleotides. In the liver, which responds to glucagon and cyclic AMP with an increase in gluconeogenesis and glycogenolysis it would appear that cyclic AMP may duplicate the changes in permeability seen with glucagon. However, it is possible to produce an increase in cyclic AMP without changes in ion transport. Moreover, increases in cyclic AMP formation in the liver in response to glucagon have not been demonstrated earlier than 30 seconds or prior to membrane potential changes. The possible direct effect of cell surface active hormones on ion flow is further suggested by the *in vitro* observations of a glucagon and epinephrine inhibition of (Na^+/K^+)-ATPase and the inhibition of this effect with insulin. These effects on (Na^+/K^+)-ATPase and Mg^{2+}-ATPase and possibly guanyl cyclase take on additional significance where there is no stimulation of cyclic AMP formation by hormones e.g. insulin, growth hormone, acetylcholine.

Although the major contribution to changes in membrane potential comes from the monovalent cations, there is ample evidence for hormone-stimulated divalent cation transport, particularly calcium. The critical roles of calcium [24, 267–269] and magnesium in intracellular and plasma membrane function have been well documented. In this regard, are the interesting observations that peptide hormones may influence calcium binding by plasma membranes [218, 303].

1.5.2 An integrated model for the mechanism of hormone action

Perspectives
In order to consider the mechanism of hormone action at the cell surface we must first consider the basal or resting state of the target cell. This is important since the perspective we wish to impart places emphasis on the plasma membrane regulation of a dynamic steady state condition of the cell. Briefly stated this condition consists of: (a) a resting electric potential reflecting ion gradients across the plasma membrane maintained by active ion pumps; (b) basal levels of cytoplasmic divalent cations, which in the case of Ca^{2+} is about three orders of magnitude lower than plasma and mitochondrial pools; (c) basal levels of cytoplasmic cyclic nucleotides; (d) transmembrane nutrient gradients e.g. glucose, amino

acids, etc.; and (e) a basal level of endogenous hormone bathing the target cells and resulting in partial occupancy of its receptors. The topographical distribution of hormone receptors is in a clustered or patched state, the degree of which, depends on the specific receptor, and target cell and possibly on submembranous microfilaments and microtubules. The mechanism of hormone action must then be considered as a transient perturbation of the above dynamic steady state of the cell manifested through the plasma membrane in response to a change in plasma hormone concentration. The energy source for this perturbation could be the substantial energy stored in the transmembrane ion gradients and electric potential. The model we outline below elaborates on the nature of this perturbation more fully.

An important contribution to the model is the assumption that hormones and neurotransmitters have a similar mode of action via their cell surface receptors. Although this point is not new, we feel it worth emphasizing since there is a wealth of information provided by the neurophysiologists and neurochemists on bioelectric properties, membrane ion fluxes, and pharmacological agents that has not been fully exploited in most previous considerations of hormone–receptor interactions. Rasmussen [267–269] has taken important steps in this direction, placing primary emphasis on cytoplasmic calcium as an additional second messenger to cyclic AMP. The assumption of similarities in the mode of action of hormone and neurotransmitters is supported by several interesting observations. Hormones like neurotransmitters have been found to cause changes in membrane electric potential with accompanying ion fluxes. Thus glucagon, like catecholamines, induces hyperpolarization of a variety of target cells with similar ion translocations (see Section 1.4). Furthermore, omission of Na^+ from rat liver perfusion medium blocked glucagon-induced glycogenolysis and gluconeogenesis without affecting the glucagon stimulation of adenylate cyclase. Action-potential-like spikes have been observed for the peptide hormone, adrenocorticotropin, acting on the adrenal gland [215]. On the other hand, the catecholamine neurotransmitters are known to stimulate adenylate cyclase at β-adrenergic receptors as do many hormones at their respective receptors. In fact, the catecholamines, representing both neurotransmitters and endocrine hormones, may exemplify the common mechanism for the two categories of cell-activating ligands. It is tempting to speculate that the similarities between glucagon and catecholamines will be mirrored by their counter-hormones, insulin and acetylcholine.

Membrane response to hormone perturbation

With the above perspectives we now outline an integrated model of the mechanism of hormone action via cell surface receptors. The following series of molecular events are postulated to occur on the indicated minimum time scale reflecting existing data. In some cases the minimum measured time may reflect experimental limitations rather than a rate-limiting membrane response. In addition, we list several important regulatory feedback control loops as part of the integrated model of hormone action.

Hormone perturbation of membrane dynamic steady state

$t = 0$	Receptor topography: partially clustered or patched; Receptors partially occupied via basal hormone level
$t \simeq$ ms–s	Hormone binding changes in response to altered plasma hormone concentration
$t \simeq$ ms–s	Membrane ion permeability changes; Ion flux; Electric potential change; Shift in Cytoplasmic Ca^{2+}, Mg^{2+}, Na^+, K^+, H^+ concentration
$t \simeq$ s	Propagated changes of state of membrane lipid and proteins (monitored spectroscopically)
$t \simeq$ s	Membrane enzyme activation or inhibition (Adenylate cyclase, ATPase, phosphodiesterase, protein kinase, guanylate cyclase?) Shift in concentration of cyclic nucleotides
$t \simeq$ s–min	Membrane transport activation; Shift in cytoplasmic concentration of sugars, amino acids, fatty acids, etc.
$t \simeq$ ms–h	Cellular biological response: metabolism, muscle contraction, nervous transmission and conduction, secretion, etc.

Regulatory feedback control loops

Hormone binding modulation via

(i) Shift in binding affinity and dissociation rate with changes in receptor occupancy,

(ii) Enzymatic degradation e.g. acetylcholinesterase, monoamine oxidase, insulinase.

Ion concentration modulation via

(i) Activation/inhibition of ATPase ion pumps,

(ii) Permeability changes due to alteration in membrane bound Ca^{2+},

(iii) Cyclic nucleotide alteration of plasma membrane and mitochondrial ion permeability.

Cyclic nucleotide modulation via

(i) Ca^{2+}, Mg^{2+}, pH, guanylnucleotide action on cyclase,
(ii) Ca^{2+} activation of phosphodiesterase.

Plausible molecular mechanisms

The clustered and patched state of hormone receptors observed for a variety of systems (Section 1.2.2) could be under the influence of cytoplasmic microfilaments and microtubules. There are almost no data on this for hormone systems at the present time other than one paper [333] reporting that cytochalasins reduce insulin and growth hormone binding to cultured lymphocytes. However, the role of microfilaments and microtubules in toxin, lectin, and immunoreceptor patching and capping has been shown to be important and it is clearly possible that the hormone receptor topography might be similarly regulated. If microfilament disrupting agents (i.e. cytochalasins) disperse hormone receptor clusters or patches, this might in turn influence hormone binding as noted by Van Obberghan *et al.* [333]. Furthermore, microtubular structures are known to be sensitive to divalent cations and nucleotides, as is the membrane cyclic AMP generating system. Thus hormone perturbation of cytoplasmic Ca^{2+} and GTP levels could alter submembranous microtubules which might then feedback to alter receptor topography, hormone binding affinity, and cyclic AMP generation. This also provides a mechanism for the hormone–receptor complex to influence the microtubule system which could then function in other cellular activities.

Other influences on receptor topography could come from lipid phase separation stimulated by cation perturbations, valency of the hormone, and degree of occupancy of the receptors. The cholera toxin inspired mobile receptor hypothesis [20, 156] would suggest that receptor clustering and patching is induced by the ligand-receptor interaction. However, the parallel effect has not been demonstrated for a hormone receptor, and there is some limited evidence suggesting the clustered state of hormone receptors is independent of hormone binding (Section 1.2.2). Most of these studies however, were performed under receptor exposure to basal levels of endogenous hormone, whereas for cholera toxin there is no endogenous ligand. Notwithstanding the latter, the time required for lateral diffusion of membrane proteins across a few microns of cell surface (minutes) (see Section 1.2.1) appears longer than measured times of stimulated ion fluxes and enzyme activation (ms–s).

There are several ways in which hormone binding and dissociation rates could be regulated as a function of receptor occupancy. Enzymatic degradation and conversion is surely involved as suggested by examples of acetylcholinesterase, monoamine oxidase, catechol-O-methyl transferase, insulinase, and other peptide 'hormonases'. Beyond this, there is evidence of a shifting binding affinity and dissociation rates in response to receptor occupancy, i.e. non-linear Scatchard plots. Although the validity of the assumptions underlying the application of Scatchard analysis to radio-labeled specific hormone binding has come into serious question recently [46, 155, 326], we will assume for the present discussion that binding affinities and dissociation rates can shift with varying receptor occupancy. There are many molecular pathways by which this could come about. Hormone-stimulated changes in the membrane ion environment and electric potential might feed back to alter receptor binding properties as has been observed for the scorpion toxin receptor [45]. Membrane enzymes or transport proteins associating with the hormone–receptor complex could change the receptor's binding properties. There is also the possibility of co-operative or allosteric transitions among receptors that are a function of receptor occupancy. These could be triggered by changes in receptor clustering size and density in response to altered ion gradients or membrane electric potential. The result would be a co-operative propagation of the shift in binding properties throughout the membrane to both occupied and unoccupied receptors. The more hormone binding, the greater the ion perturbation, and the greater the shift toward decreased binding affinities. Another co-operative mechanism could involve direct interaction of occupied sites on a lattice, the frequency of such interactions obviously increasing with site occupancy. This mechanism might be modeled by a two dimensional Ising lattice [138] and might allow for transformations between different affinity states of the receptor. If these affinity states depended on whether the occupied receptor were in a cluster or freely dispersed in the membrane, then the transformations could also correspond to a two dimensional phase transition between free and clustered states of these receptors.

Several mechanisms could contribute to the control of membrane-bound enzyme activity by hormones. Since mobility has not yet been demonstrated for a hormone–receptor complex, it may be that the basal mobility of the effector, e.g. adenylate cyclase, in the membrane plane would allow it to migrate to and be recognized by the less mobile hormone–receptor complex. Such an altered mobility could be the consequence of altered membrane microviscosity. It has been reported that the Hill coefficients of the erythrocyte (Na^+/K^+)-ATPase and acetylcholinesterase response to insulin is a function of membrane microviscosity [213]. Although the mechanism for a proposed change in microviscosity is not known, it may be one of the consequences of an altered ionic medium as reflected in changes in membrane potential. Calcium and magnesium and to a lesser extent, monovalent cations are known to have significant effects on lipid phase transitions and separations [157, 246]. Moreover, such phase transitions may alter the availability of specific lipids for specific membrane-bound enzymes e.g. (Na^+/K^+)-ATPase, adenylate cyclase [52].

In addition to the effect of lipids on membrane-bound enzyme activity there are direct effects of ions on these enzymes. Of particular interest are those ions which are translocated in association with hormone–membrane binding. Thus calcium concentrations [269] may regulate adenylate cyclase, ATPase, and phosphodiesterase activities. Adenylate cyclase activity is additionally influenced by magnesium ion and pH, both of which may be altered in response to hormones (see Section 1.4). The activity of (Na^+/K^+)-ATPase would similarly be influenced by the concentrations of these two cations the transport of which is stimulated by hormones (see Section 1.4).

Ion translocation and change in membrane potential are among the first consequences of hormone and neurotransmitter binding to specific receptor sites (see Section 1.4). If indeed these changes are immediate consequences of binding, what mechanisms could account for this? It may be that the hormone receptor is identical with or closely associated with the ionophore as has been demonstrated with acetylcholine and electroplax membrane [133, 225, 226]. Perhaps the association of insulin receptors and intramembranous particles noted [244] by freeze-fracture electron microscopy is an indication of the association of the receptor with an ionophore. Alternatively, association of the hormone with its receptor may result in changes in receptor packing or clustering to modify ion channels, reminiscent of alamethacin ion channels [8]. In either case, hormone binding could produce a conformational change with stimulated ion flow. Hormones have also been shown to alter membrane calcium binding [218, 303, 330], which in turn could affect membrane ion permeability [283]. Perhaps the hormone interacts with one or more enzymes which in turn affect ion translocation. (Na^+/K^+)-ATPase has been demonstrated to be affected *in vitro* by peptide hormones. Stimulation of adenylate cyclase, production of cyclic AMP, protein kinase activation and membrane phosphorylation might also lead to ion translocation. Cyclic AMP can reproduce some of the effects of hormones on ion translocation. This latter mechanism would not account for those hormones and neurotransmitters which affect ion translocation without the mediation of cyclic AMP e.g. insulin, acetylcholine, growth hormone.

From spectoscopic studies performed on cell surface active hormones and neurotransmitter interaction with specific target cell membranes it is apparent that there are widespread changes in various membrane components. This would be consistent with changes in hormone binding, ion translocation and multiple enzyme stimulation. It may be that these widespread membrane changes are a consequence of ion translocation, viscosity changes or membrane phosphorylation due to adenylate cyclase and protein kinase activation.

1.5.3 Comments on the model

There are several aspects of the integrated model of hormone action outlined above (Section 1.5.2) that are worth noting: First, in contrast to the classical view of hormone action mediated through adenylate cyclase-generated cyclic AMP, it proposes multiple membrane enzyme activation with a variety of second messengers, including both ions and nucleotides. Second, we have emphasized a common mode of action for hormones and neurotransmitters thereby stressing the importance of transmembrane ion gradients, ion fluxes, and electric potential. We have not confined our view in this regard to calcium, but believe that Na^+, K^+, Mg^{2+} and H^+, as well as anions may be directly involved. The millisecond time scale of ion fluxes and electric potential changes across excitable membrane suggest the possibility that these may be among the earliest events following hormone binding which could trigger subsequent membrane-mediated responses to the hormone. A propagated change in ion permeability might induce the spectroscopic changes of state observed in several target membranes. Third, we have incorporated into the model recent information on the molecular topography of hormone receptors in a fluid membrane. Fourth, we have provided the model with regulatory feedback loops controlling hormone binding, second messenger levels, and ion fluxes. Finally, we have stressed a certain perspective which recognizes the dynamic steady state of the target cell under basal or resting conditions and considers the hormone action as a transient perturbation or shift in this steady state.

We have borrowed liberally from the neurochemistry literature as well as from various models of hormone action, some of which emphasize binding [75, 164] or enzyme activation [20, 127], while others emphasize the role of calcium [24, 267–269] or receptor mobility in a fluid membrane [19, 20, 72, 156, 219]. To these we have added recent evidence on membrane mobility and receptor topography, membrane spectroscopic responses, and the perspective indicated in Section 1.5.2 in formulating what we have called an 'integrated' model of hormone action. While there is the danger of a model so general that it could account for anything, we have attempted to provide an overall view which focuses on the interelationships among certain key membrane regulatory phenomena and which can serve as a useful guide to future experiments. Thus, it would be interesting to measure the early kinetics of ion permeability changes induced by hormones as compared with early enzyme activation, on the ms–s time scale. Similarly it would be interesting to see which membrane events can be bypassed or blocked (possibly via pharmacological agents) and what the consequences are for other membrane actions and the final biological response. For example, experiments on the effects of ionophores or local anesthetic ion blockers on enzyme activation and the biological response could be extended and might help elucidate the sequence of events following hormone binding. Future ultrastructural mapping of hormone receptors could profitably take into account the effects of receptor occupancy and microtubule and microfilament structures on the state of receptor clustering and patching. These are but a few of the unanswered questions posed by the above model. It will have served a useful purpose if it stimulates further experiments and ideas on the molecular mechanism by which cell surface active hormones regulate cellular life.

REFERENCES

1. Abramson, M.B. (1971), *The Chemistry of Biosurfaces,* 1, (Hair, M.L., ed.), Marcel Dekker, New York, p. 45.
2. Aizono, Y., Roberts, J.E., Sonenberg, M. and Swislocki, N.I. (1974), *Arch. Biochem. Biophys.*, **163**, 634–643.
3. Alberty, R.A. and Hammes, G.G. (1958), *J. Phys. Chem.*, **62**, 154–159.
4. Albuquerque, E.X., Barnard, E.A., Porter, C.A. and Warnick, J.E. (1974), *Proc. natn. Acad. Sci., U.S.A.*, **71**, 2818–2822.
5. Alexander, R.W., Williams, L.T. and Lefkowitz, R.J. (1975), *Proc. natn. Acad. Sci., U.S.A.*, **72**, 1564–1568.

6. Allen, J.E. and Rasmussen, H. (1971), *Science,* **174**, 512–514.
7. Allen, J.E. and Valeri, C.R. (1974), *Arch. Int. Med.,* **133**, 86–96.
8. Allison, A.C. (1974), *Control of Proliferation in Animal Cells,* (Clarkson, B. and Baserga, R., eds.), Cold Spring Harbor Laboratory, pp. 447–459.
9. Anderson, M.J., and Cohen, M.W. (1974), *J. Physiol.,* **237**, 385–400.
10. Anderson, R.G.W., Goldstein, J.L. and Brown, M.S. (1976), *Proc. natn. Acad. Sci., U.S.A.,* **73**, 2434–2438.
11. Barden, N. and Labrie, F. (1973), *J. biol. Chem.,* **248**, 7601–7606.
12. Barnabei, O., Leghissa, G. and Tomasi, V. (1974), *Biochim. biophys. Acta,* **362**, 316–325.
13. Barnett, R.E., Scott, R.E., Furcht, L.T. and Kersey, J.H. (1974), *Nature,* **249**, 465–466.
14. Bashford, C.L., Harrison, S.J., Radda, G.K. and Mehdi, Q. (1975), *Biochem. J.,* **146**, 473–479.
15. Batt, R. and McKenzie, J.M. (1976), *Am. J. Physiol.,* **231**, 52–55.
16. Benda, P., Tsuji, S., Daussant, T. and Changeux, J-P. (1969), *Nature,* **225**. 1149–1150.
17. Benjamin, W.B. and Singer, I. (1975), *Biochemistry,* **14**, 3301–3309.
18. Bennett, H.S. (1969), *Handbook of Molecular Cytology.,* (Lima-de Faria, ed.), North Holland Pub., Amsterdam, pp. 1294–1319.
19. Bennett, V., Craig, S., Hollenberg, M.D., O'Keefe, E., Sahyoun, N. and Cuatrecasas, P. (1976), *J. supramolec. Struct.,* **4**, 99–120.
20. Bennett, V., O'Keefe, E. and Cuatrecasas, P. (1975), *Proc. natn. Acad. Sci., U.S.A.,* **72**, 33–37.
21. Berlin, R.D. (1975), *Ann. N.Y. Acad. Sci.,* **253**, 445–454.
22. Berlin, R.D., Oliver, J.M., Ukena, T.E. and Yin, H.H. (1974), *Nature,* **247**, 45–46.
23. Berlin, R.D. and Ukena, T.E. (1972), *Nature (New Biol.),* **238**, 120–122.
24. Berridge, M.J. (1975), *Adv. Cyclic Nucleotide Res.,* **6**, 1–99.
25. Biener, J., Jansen, F.K. and Brandenburg, D. (1973), *Diabetologia,* **9**, 53–55.
26. Bieri, V.G., Wallach, D.F.H. and Lin, P.S. (1974), *Proc. natn. Acad. Sci., U.S.A.,* **71**, 4797–4801.
27. Bilezikian, J.P. and Aurbach, G.D. (1973), *J. biol. Chem.,* **248**, 5577–5583.
28. Bilezikian, J.P. and Aurbach, G.D. (1973), *J. biol. Chem.,* **248**, 5584–5589.
29. Blasie, J.K. (1972), *Biophysical J.,* **12**, 191–204.
30. Bloom, F.E. and Barrnett, R.J. (1966), *J. Cell Biol.,* **29**, 475–495.
31. Bourgeois, J.P., Ryter, A., Menez, A., Fromageot, P., Boquet, P. and Changeux, J-P. (1972), *FEBS Letters.,* **25**, 127–133.
32. Bourgeois, J.P., Tsugi, S., Boquet, P., Pillot, J., Ryter, A., and Changeux, J-P. (1971), *FEBS Letters.,* **16**, 92–94.
33. Branton, D. and Park, R.B. (1968), *Biological Membrane Structure,* Little Brown and Company, Boston.

34. Bretscher, M.S. (1972), *J. mol. Biol.*, **71**, 523–528.
35. Bretscher, M.S. (1973), *Science*, **181**, 622–629.
36. Bretscher, M.S. and Raff, M.C. (1975), *Nature*, **258**, 43–49.
37. Brewer, G.J. (1974), *Biochemistry*, **13**, 5038–5045.
38. Brewer, G.J. (1976), *Biochemistry*, **15**, 1387–1392.
39. Brisson, A.D., Scandella, C.J., Bienvenue, A., Devaux, P.F., Cohen, J.B. and Changeux, J-P. (1975), *Proc. natn. Acad. Sci., U.S.A.*, **72**, 1087–1091.
40. Brown, M.S. and Goldstein, J.L. (1976), *Science*, **191**, 150–154.
41. Brown, S.S. and Revel, J-P. (1976), *J. Cell Biol.*, **68**, 629–641.
42. Butcher, R.W., Croford, O.B., Gammeltoff, S., Gliemann, J., Gavin III, J.R., Goldfine, I.D., Kahn, C.R., Rodbell, M., Roth, J., Jarrett, L., Larner, J., Lefkowitz, R.J., Levine, R. and Marinetti, G.V. (1973), *Science*, **182**, 396–397.
43. Butterfield, D.A., Whisnant, C.C. and Chestnut, D.B. (1976), *Biochim. biophys. Acta*, **426**, 697–702.
44. Cadenhead, D.A., Kellner, B.M.J. and Muller-Landau, F. (1975), *Biochim. biophys. Acta*, **382**, 253–259.
45. Catterall, W.A., Ray, R. and Morrow, C.S. (1976), *Proc. natn. Acad. Sci., U.S.A.*, **73**, 2682–2686.
46. Chang, K-J., Jacobs, S. and Cuatrecasas, P. (1975), *Biochim. biophys. Acta*, **406**, 294–303.
47. Chang, K-J., Marcus, N.A. and Cuatrecasas, P. (1974), *J. biol. Chem.*, **249**, 6854–6865.
48. Chapman, D. (1975), *Quart. Rev. Biophys.*, **8**, 185–235.
49. Cohen, J.B. and Changeux, J-P. (1973), *Biochemistry*, **12**, 4855–4864.
50. Cohen, J.B. and Changeux, J-P. (1975), *Ann. Rev. Pharmacol.*, **15**, 83–103.
51. Cohen, J.B., Weber, M., Huchet, M. and Changeux, J-P. (1972), *FEBS Letters*, **26**, 43–47.
52. Coleman, R. (1973), *Biochim. biophys. Acta*, **300**, 1–30.
53. Collander, R. and Bärlund, H. (1933), *Acta Bot. Fenn.*, **11**, 1–117.
54. Cowgill, R.W. (1970), *Biochem. biophys. Acta*, **200**, 18–25.
55. Craig, S.W. and Cuatrecasas, P. (1975), *Proc. natn. Acad. Sci., U.S.A.*, **72**, 3844–3848.
56. Cramer, W.A. and Keenan, T.W. (1974), *Biochem. biophys. Res. Comm.*, **56**, 60–67.
57. Cramer, W.A. and Phillips, S.K. (1970), *J. Bact.*, **104**, 819–825.
58. Cramer, W.A., Phillips, S.K. and Keenan, T.W. (1973), *Biochemistry*, **12**, 1177–1181.
59. Cuatrecasas, P. (1969), *Proc. natn. Acad. Sci., U.S.A.*, **63**, 450–457.
60. Cuatrecasas, P. (1973), *Biochemistry*, **12**, 3547–3558.
61. Cuatrecasas, P. (1974), *Ann. Rev. Biochem.*, **43**, 169–214.
62. Cuatrecasas, P., Parikh, I. and Hollenberg, M.D. (1973), *Biochemistry*, **12**, 4253–4264.

63. Cuatrecasas, P., Tell, G.P.E., Sica, V., Parikh, I. and Chang, K-J. (1974), *Nature,* **247**, 92–97.
64. Dambach, G. and Friedmann, N. (1974), *Biochim. biophys. Acta,* **332**, 374–386.
65. Dambach, G. and Friedmann, N. (1974), *Biochim. biophys. Acta.* **367**, 366–370.
66. Dandliker, W.B. and de Saussure, V.A. (1971), *The Chemistry of Biosurfaces,* Vol. 1, (Hair, M.L., ed.), Marcel Dekker, New York, p. 1.
67. Danielli, J.F. and Davson, H. (1935), *J. Cell comp. Physiol.,* **5**, 495–508.
68. Daniels, M. and Vogel, Z. (1975), *Nature,* **254**, 339–341.
69. Davidson, M.B., Van Herle, A.J. and Gerschenson, L.E. (1973), *Endocrinology,* **92**, 1442–1446.
70. Davis, D.G. (1972), *Biochem. biophys. Res. Commun.,* **49**, 1492–1497.
71. Deamer, D.W. and Branton, D. (1967), *Science,* **158**, 655–657.
72. DeHaën, C. (1976), *J. Theor. Biol.,* **58**, 383–400.
73. De Lorenzo, R.J., Walton, K.G., Curran, P.F. and Greengard, P. (1973), *Proc. natn. Acad. Sci., U.S.A.,* **70**, 880–884.
74. De Meyts, P. (1976), *J. supramolec. Struct.,* **4**, 241–258.
75. De Meyts, P., Bianco, A.R., Roth, J. (1976), *J. biol. Chem.,* **251**, 1877–1888.
76. De Petris, S. (1974), *Nature,* **250**, 54–56.
77. De Petris, S. (1975), *J. Cell Biol.,* **65**, 123–146.
78. De Petris, S., Raff, M.C. and Mallucci, L. (1973), *Nature (New Biol.),* **244**, 275–278.
79. Devaux, P. and McConnell, H.M. (1972), *J. Am. Chem. Soc.,* **94**, 4475–4481.
80. Devynck, M.A., Pernollet, M-G. and Meyer, P. (1973), *Nature (New Biol.),* **245**, 55–58.
81. Donner, D.B. and Sonenberg, M. (unpublished).
82. Duguid, J.R. and Raftery, M.A. (1973), *Biochemistry,* **12**, 3593–3597.
83. DuPont, Y., Cohen, J.B. and Changeux, J-P. (1974), *FEBS Letters,* **40**, 130–133.
84. Edelman, G.M. (1976), *Science,* **192**, 218–226.
85. Edelman, G.M., Yahara, I. and Wang, J.L. (1973), *Proc. natn. Acad. Sci., U.S.A.,* **70**, 1442–1446.
86. Edidin, M. (1974), *Ann. Rev. Biophys. Bioeng.,* **3**, 179–201.
87. Edidin, M. and Fambrough, D. (1973), *J. Cell Biol.,* **57**, 27–37.
88. Einstein, A. (1956), *Investigations on the Theory of the Brownian Motion.,* Dover, New York, pp. 66–85.
89. Ekman, A., Manninen, V. and Salminen, S. (1969), *Acta Physiol. Scand.,* **75**, 333–344.
90. Eldefrawi, M.E., Eldefrawi, A.T. and Wilson, D.B. (1975), *Biochemistry,* **14**, 4304–4310.

91. Elgsaeter, A. and Branton, D. (1974), *J. Cell. Biol.*, **63**, 1018–1036.
92. Epstein, G.H. and Whittam, R. (1966), *Biochem. J.*, **99**, 232–238.
93. Evans, R.B., Morhenn, V., Jones, A.L. and Tomkins, G.M. (1974), *J. Cell Biol.*, **61**, 95–106.
94. Exton, J.H., Robison, G.A., Sutherland, E.W. and Park, C.R. (1971), *J. biol. Chem.*, **246**, 6166–6177.
95. Farias, R.N., Bloj, B., Morero, R.D., Sineriz, F. and Trucco, R.E. (1975), *Biochim. Biophys. Acta,* **415**, 231–251.
96. Fasman, G.D., Bodenheimer, E. and Pesce, A. (1966), *J. biol. Chem.*, **241**, 916–923.
97. Fasman, G.D., Norland, K. and Pesce, A. (1964), *Biopolym. Symp.*, **1**, 325–331.
98. Feingold, D.S. (1970), *J. Memb. Biol.*, **3**, 372–386.
99. Feitelson, J. (1969), *Photochem. and Photobiol.*, **9**, 401–410.
100. Fertuck, H.C. and Salpeter, M.M. (1976), *J. Cell Biol.*, **69**, 144–158.
101. Field, J.B., Bloom, G., Kerins, M.E., Chayoth, R. and Zor, U. (1975), *J. biol. Chem.*, **250**, 4903–4910.
102. Fields, K.L. and Luria, S.E. (1969), *J. Bact.*, **97**, 57–63.
103. Fields, K.L. and Luria, S.E. (1969), *J. Bact.*, **97**, 64–77.
104. Finch, E.D. and Schneider, A.S. (1975), *Biochim. biophys. Acta,* **406**, 146–154.
105. Finean, J.B. (1969), *Quart. Rev. Biophys.*, **2**, 1–23.
106. Fortes, P.A.G. and Ellory, J.C. (1975), *Biochim. biophys. Acta,* **413**, 65–78.
107. Fortes, P.A.G. and Hoffman, J.F. (1974), *J. Memb., Biol.*, **16**, 79–100.
108. Friedmann, N. (1972), *Biochim. biophys. Acta,* **274**, 214–225.
109. Friedmann, N. (1976), *Biochim. biophys. Acta,* **428**, 495–508.
110. Friedmann, N. and Dambach, G. (1973), *Biochim. biophys. Acta,* **307**, 399–403.
111. Friedmann, N. and Rasmussen, H. (1970), *Biochim. biophys. Acta,* **222**, 41–52.
112. Friedmann, N., Somlyo, A.V. and Somlyo, A.P. (1971), *Science,* **171**, 400–402.
113. Frye, L.D. and Edidin, M. (1970), *J. Cell Sci.*, **7**, 319–335.
114. Fuchs, P., Parola, A., Robbins, P.W. and Blout, E.R. (1975), *Proc. natn. Acad. Sci., U.S.A.*, **72**, 3351–3354.
115. Gautron, J. (1970), *C.R. Acad. Sci. Paris,* Ser. D., **271**, 714–717.
116. Gavin, J.R. III, Roth, J., Jen, P. and Freychet, P. (1972), *Proc. natn. Acad. Sci., U.S.A.*, **69**, 747–751.
117. Gill, D.M. and King, C.A. (1975), *J. biol. Chem.*, **250**, 6424–6432.
118. Ginsberg, B.H., Kahn, C.R. and Roth, J. (1976), *Biochim. biophys. Acta,* **443**, 227–242.
119. Gitter- Amir, A., Rosenheck, K. and Schneider, A.S. (1976), *Biochemistry,* **15**, 3131–3137.

120. Glynn, I.M. and Warner, A.E. (1972), *Br. J. Pharmacol.*, **44**, 271–278.
121. Godfraind, T. (1976), *J. Physiol.*, **260**, 21–35.
122. Goldstein, J.L. and Brown, M.S. *Current Topics in Cellular Regulation*, Vol. 11, (Horecker, B.L. and Stadtman, E.R., eds.), Academic Press, New York, (in press).
123. Gorter, E. and Grendel, F. (1925), *J. exp. Med.*, **41**, 439–443.
124. Gunther, G.R., Wang, J.L., Yahara, I., Cunningham, B.A. and Edelman, G.M. (1973), *Proc. natn. Acad. Sci., U.S.A.*, **70**, 1012–1016.
125. Haber, E. and Wrenn, S. (1976), *Physiol. Rev.*, **56**, 317–338.
126. Hadden, J.W., Hadden, E.M., Wilson, E.E. and Good, R.A. (1972), *Nature (New Biol.)*, **235**, 174–176.
127. Hammes, G.G. and Rodbell, M. (1976), *Proc. natn. Acad. Sci., U.S.A.*, **73**, 1189–1192.
128. Hartzell, H.C. and Fambrough, D.M. (1972), *J. Gen. Physiol.*, **60**, 248–262.
129. Haylett, D.G. and Jenkinson, D.H. (1969), *Nature*, **224**, 80–81.
130. Haylett, D.G. and Jenkinson, D.H. (1972), *J. Physiol.*, **225**, 721–750.
131. Haylett, D.G. and Jenkinson, D.H. (1972), *J. Physiol.*, **225**, 751–772.
132. Haynes, D.H. and Staerk, H. (1974), *J. memb. Biol.*, **17**, 313–340.
133. Hazelbauer, G.L. and Changeux, J-P. (1974), *Proc. natn. Acad. Sci., U.S.A.*, **71**, 1479–1483.
134. Helgerson, S.L., Cramer, W.A., Harris, J.M. and Lytle, F. (1974), *Biochemistry*, **13**, 3057–3061.
135. Helmreich, E.J.M., Zenner, H.P., Pfeuffer, T. and Cori, C.F. (1976), *Current Topics in Cell Regulation*, **10**, 41–87.
136. Herschman, H.R. and Helinski, D.R. (1967), *J. biol. Chem.*, **242**, 5360–5368.
137. Hershko, A, Mamont, P., Shields, R. and Tomkins, G.M. (1971), *Nature (New Biol.)*, **232**, 206–211.
138. Hill, T.L. (1960), *An Introduction to Statistical Thermodynamics*, Chap. 14, Addison Wesley, Reading, Mass.
139. Höber, R. (1945), *Physical Chemistry of Cells and Tissues*, Chap. 10, 15, The Blakiston Co., Philadelphia.
140. Hohenwallner, W. and Ludescher, E. (1970), *Clin. chim., Acta*, **28**, 255–258.
141. Holmgren, J. and Lönnroth, I. (1975), *FEBS Letters*, **55**, 138–142.
142. Holmgren, J., Lönnroth, I., Mansson, J.E. and Svennerholm, L. (1975), *Proc. natn. Acad. Sci., U.S.A.*, **72**, 2520–2524.
143. Holmgren, J., Lönnroth, I. and Svennerholm, L. (1973), *Infect. Immun.*, **8**, 208–214.
144. Hopfer, U., Lehninger, A.L. and Thompson, T.E. (1968), *Proc. natn. Acad. Sci., U.S.A.*, **59**, 484–490.
145. Horowitz, A.F., Horsley, W.J. and Klein, M.P. (1972), *Proc. natn. Acad. Sci., U.S.A.*, **69**, 590–593.

146. Howell, S.L. and Whitfield, M. (1972), *J. Histochem. Cytochem.*, **20**, 873–879.

147. Hsueh, A.J.W. and Catt, K.J. (private communication).

148. Hsueh, A.J.W., Dufau, M.L., Katz, S.I. and Catt, K.J. (1976), *Nature*, **261**, 710–712.

149. Huestis, W. (1976), *J. Supramolec. Struct.*, **4**,355–365.

150. Huestis, W.H. and McConnell, H.M. (1974), *Biochem. biophys. Res. Commun.*, **57**, 726–732.

151. Illiano, G., Tell, P.E., Siegel, M.I. and Cuatrecasas, P. (1973), *Proc. natn. Acad. Sci., U.S.A.*, **70**, 2443–2447.

152. Imae, T., Fasman, G.D., Hinkle, P.M. and Tashjian, A.H. Jr. (1975), *Biochem. biophys. Res. Comm.*, **62**, 923–932.

153. Israel, M., Gautron, J. and Lesbats, B. (1970), *J. Neurochem.*, **17**, 1441–1450.

154. Jacob, F., Siminovitch, L. and Wollman, E. (1952), *Ann. Inst. Pasteur*, **83**, 295–315.

155. Jacobs, S., Chang, K-J. and Cuatrecasas, P. (1975), *Biochem. biophys. Res. Commun.*, **66**, 687–692.

156. Jacobs, S. and Cuatrecasas, P. (1976), *Biochim. biophys. Acta*, **433**, 482–495.

157. Jacobson, K. and Papahadjopoulos, D. (1975), *Biochemistry*, **14**, 152–161.

158. James, R., Branton, D., Wisnieki, B. and Keith, A. (1972), *J. Supramolec. Struct.*, **1**, 38–49.

159. Jarett, L. and Smith, R.M., (1974), *J. biol. Chem.*, **249**, 7024–7031.

160. Jarett, L. and Smith, R.M. (1975), *Proc. natn. Acad. Sci., U.S.A.*, **72**, 3526–3530.

161. Jarett, L. and Smith, R.M. (1976), *Abstract American Diabetes Meeting*, **4**, p. 321.

162. Johnson, L.W., Hughes, M.E. and Zilversmit, D.B. (1975), *Biochim. biophys. Acta*, **375**, 176–185.

163. Kachadorian, W.A., Wade, J.B. and DiScala, V.A. (1975), *Science*, **190**, 67–69.

164. Kahn, C.R. (1976), *J. Cell Biol.*, **70**, 261–286.

165. Karlin, A. (1967), *Proc. natn. Acad. Sci., U.S.A.*, **58**, 1162–1167.

166. Karlin, A. (1974), *Life Sci.*, **14**, 1385–1415.

167. Kasai, M., Changeux, J-P. and Monnerie, L. (1969), *Biochem. biophys. Res. Commun.*, **36**, 420–427.

168. Katzen, H.M. and Vlahakes, G.J. (1973), *Science*, **179**, 1142–1143.

169. Ke, B. (1965), *Archs. Biochem. Biophys.*, **112**, 554–561.

170. Keely, S.L., Corbin, J.D. and Park, C.R. (1975), *J. biol. Chem.*, **250**, 4832–4840.

171. Kim, S.K. (1976), *J. Supramolec. Struct.*, **4**, 185–197.

172. Kimberg, D.V., Field, M., Johnson, J., Henderson, A. and Gershon, E. (1971), *J. clin. Invest.*, **50**, 1218–1230.
173. King, C.A. and Van Heyningen, W.E. (1973), *J. Infect. Dis.*, **127**, 639–647.
174. Kleeman, W., Grant, C.W.M. and McConnell, H.M. (1974), *J. Supramolec. Struct.*, **2**, 609–616.
175. Kleeman, W. and McConnell, H.M. (1974), *Biochim. biophys. Acta*, **345**, 220–230.
176. Kornberg, R.D. and McConnell, H.M. (1971), *Biochemistry*, **10**, 1111–1120.
177. Kram, R., Mamont, P. and Tomkins, G.M. (1973), *Proc. natn. Acad. Sci., U.S.A.*, **70**, 1432–1436.
178. Krishnan, K.S. and Balaram, P. (1975), *FEBS Letters*, **60**, 419–422.
179. Kury, P.G. and McConnell, H.M. (1975), *Biochemistry*, **14**, 2798–2803.
180. Kury, P.G., Ramwell, P.W. and McConnell, H.M. (1974), *Biochem. biophys. Res. Commun.*, **56**, 478–483.
181. Ladbrooke, B.D. and Chapman, D. (1969), *Chem. Phys. Lipids*, **3**, 304–356.
182. Lau, C. and Richards, F.M. (1976), *Biochemistry*, **15**, 666–671.
183. Lee, A.G., Birdsall, N.J.M. and Metcalfe, J.C. (1973), *Biochemistry*, **12**, 1650–1659.
184. Lefkowitz, R.J., Limbird, L.E., Mukherjee, C. and Caron, M.G. (1976), *Biochim. biophys. Acta*, **457**, 1–39.
185. Lefkowitz, R.J., Mukherjee, C., Coverstone, M. and Caron, M.G. (1974), *Biochem. biophys. Res. Commun.*, **60**, 703–709.
186. Lemay, A., and Jarett, L. (1975), *J. Cell Biol.*, **65**, 39–50.
187. Lenard, J. and Singer, S.J. (1966), *Proc. natn. Acad. Sci., U.S.A.*, **56**, 1828–1835.
188. Lesslauer, W., Cain, J.E. and Blasie, J.K. (1972), *Proc. natn. Acad. Sci., U.S.A.*, **69**, 1499–1503.
189. Levison, S.A., Dandliker, W.B., Brawn, R.J. and Vanderlaan, W.P. (1976), *Endocrinology*, **99**, 1129–1143.
190. Levitzki, A. (1974), *J. Theor. Biol.*, **44**, 367–372.
191. Levitzki, A., Atlas, D. and Steer, M.L. (1974), *Proc. natn. Acad. Sci., U.S.A.*, **71**, 2773–2776.
192. Levitzki, A., Segel, L.A. and Steer, M.L. (1975), *J. molec. Biol.*, **91**, 125–130.
193. Levitzki, A., Sevilia, N., Atlas, D. and Steer, M.L. (1975), *J. molec. Biol.*, **97**, 35–53.
194. Li, C.H. (1956), *Adv. Protein Chem.*, **11**, 121–122.
195. Lin, M.C., Salomon, Y., Rendell, M. and Rodbell, M. (1975), *J. biol. Chem.*, **250**, 4246–4252.
196. Loor, F., Forni, L. and Pernis, B. (1972), *Eur. J. Immunol.*, **2**, 203–212.
197. Lönnroth, I. and Holmgren, J. (1973), *J. Gen. Microbiol.*, **76**, 417–427.
198. LoSpalluto, J.J. and Finkelstein, R.A. (1972), *Biochim. biophys. Acta*, **257**, 158–166.

199. Lostroh, A.J. and Krahl, M.E. (1973), *Biochim. biophys. Acta,* **291**, 260–268.
200. Loten, E.G. and Sneyd, J.G.T. (1970), *Biochem. J.,* **120**, 187–193.
201. Ludescher, E. and Hohenwallner, W. (1969), *Clin. chim. Acta,* **24**, 423–430.
202. Luly, P., Barnabei, O. and Tria, E. (1972), *Biochim. biophys. Acta,* **282**, 447–452.
203. Luria, S.E. (1973), *Bacterial Membranes and Walls,* (Leive, L., ed.), Marcel Dekker, New York, pp. 293–320.
204. Maeda, A. and Nomura, M. (1966), *J. Bact.,* **91**, 685–694.
205. Maguire, M.E., Goldmann, P.H. and Gilman, A.G. (1974), *Molec. Pharmacol.,* **10**, 563–581.
206. Malchoff, C.D. and Marinetti, G.V. (1976), *Biochim. biophys. Acta,* **436**, 45–52.
207. Mallette, L.E., Neblett, M., Exton, J.H. and Langan, T.A., *J. biol. Chem.,* **248**, 6289–6291.
208. Manery, J.F. (1966), *Fed. Proc.,* **25**, 1804–1810.
209. Manery, J.F. (1969), *Mineral Metabolism, An Advanced Treatise III.,* (Comar, C. and Bronner, F. eds.), Academic Press, New York, pp. 405–452.
210. Manganiello, V. and Vaughan, M. (1973), *J. biol. Chem.,* **248**, 7164–7170.
211. Manninen, V. (1970), *Acta physiol. scand.,* **80**, Suppl. 355, 1–76.
212. Martinez, J.R., Quissell, D.O. and Giles, M. (1976), *J. Pharmac. exp. Ther.,* **198**, 385–394.
213. Massa, E.M., Morero, R.D., Bloj, B. and Farias, R.N. (1975), *Biochem. biophys. Res. Commun.,* **66**, 115–122.
214. Matthews, E.K., Petersen, O.H. and Williams, J.A. (1973), *J. Physiol.,* **234**, 689–701.
215. Matthews, E.K. and Saffran, M. (1973), *J. Physiol.,* **234**, 43–64.
216. Mayberry, H.E. and Waddell, W.J. (1974), *The Anatomical Record,* **179**, 107–114.
217. Mayr–Harting, A. (1964), *J. Path. Bact.,* **87**, 255–266.
218. McDonald, J.M., Bruns, D.E. and Jarett, L. (1976), *Proc. natn. Acad. Sci., U.S.A.,* **73**, 1542–1546.
219. McGuire, R.F. and Barber, R. (1976), *J. Supramolec. Struct.,* **4**, 259–269.
220. McNamee, M.G. and McConnell, H.M. (1973), *Biochemistry,* **12**, 2951–2958.
221. Medveczky, N. and Rosenberg, H. (1971), *Biochim. biophys. Acta,* **241**, 494–506.
222. Mehdi, S.Q. and Nussey, S.S. (1975), *Biochem. J.,* **145**, 105–111.
223. Melamed, E., Lahav, M. and Atlas, D. (1976), *Nature,* **261**, 420–422.
224. Meyers, M.B. and Swislocki, N.I. (1974), *Archs. Biochem. Biophys.,* **164**, 544–550.

225. Michaelson, D.M. and Raftery, M.A. (1974), *Proc. natn. Acad. Sci., U.S.A.*, **71**, 4768–4772.
226. Michaelson, D., Vandlen, R., Bode, J., Moody, T., Schmidt, J. and Raftery, M.A. (1974), *Archs. Biochem. Biophys.*, **165**, 796–804.
227. Miledi, R., Molinoff, P. and Potter, L.T. (1971), *Nature*, **229**, 554–557.
228. Moss, J., Fishman, P.H., Manganiello, V.C., Vaughan, M. and Brady, R.O. (1976), *Proc. natn. Acad. Sci., U.S.A.*, **73**, 1034–1037.
229. Mukherjee, C., Caron, M.G., Coverstone, M. and Lefkowitz, R.J. (1975), *J. biol. Chem.*, **250**, 4869–4876.
230. Mullin, B.R., Aloj, S.M., Fishman, P.H., Lee, G., Kohn, L.D. and Brady, R.O. (1976), *Proc. natn. Acad. Sci., U.S.A.*, **73**, 1679–1683.
231. Mullin, B.R., Fishman, P.H., Lee, G., Aloj, S.M., Ledley, F.D., Winand, R.J., Kohn, L.D. and Brady, R.O. (1976), *Proc. natn. Acad. Sci., U.S.A.*, **73**, 842–846.
232. Nachbar, M.S., Oppenheim, J.D. and Aull, F. (1974), *Am. J. med. Sci.*, **268**, 122–138.
233. Nakahara, T. and Birnbaumer, L. (1974), *J. biol. Chem.*, **249**, 7886–7891.
234. Neumann, E. and Katchalsky, A. (1972), *Proc. natn. Acad. Sci., U.S.A.*, **69**, 993–997.
235. Nicolson, G.L. (1973), *J. Cell Biol.*, **57**, 373–387.
236. Nicolson, G.L. (1976), *Biochim. biophys. Acta*, **457**, 57–108.
237. Nicolson, G.L. and Singer, S.J. (1974), *J. Cell Biol.*, **60**, 236–248.
238. Nishiyama, A. and Peterson, O.H. (1974), *J. Physiol.*, **242**, 173–188.
239. Nomura, M. (1963), *Cold Spring Harbor Symp. Quant. Biol.*, **28**, 315–324.
240. Nomura, M. (1967), *Ann. Rev. Microbiol.*, **21**, 257–283.
241. Northrop, G. (1968), *J. Pharmac. exp. Ther.*, **159**, 22–28.
242. Oesterhalt, D. and Stoeckenius, W. (1971), *Nature (New Biol.)*, **233**, 149–152.
243. Oliver, J.M., Ukena, T.E. and Berlin, R.D. (1974), *Proc. natn. Acad. Sci., U.S.A.*, **71**, 394–398.
244. Orci, L., Rufener, C., Malaisse-Lagae, F., Blondel, B., Amherdt, M., Bataille, D., Freychet, P. and Perrelet, A. (1975), *Israel J. med. Sci.*, **11**, 639–655.
245. Oski, F.A., Root, A. and Winegrad, A.I. (1967), *Nature*, **215**, 81–82.
246. Papahadjopoulos, D. and Poste, G. (1975), *Biophys. J.*, **15**, 945–948.
247. Perlmann, G.E. (1964), *Biopolym. Symp.*, **1**, 383–387.
248. Peters, R., Peters, J., Tews, K.H. and Bähr, W. (1974), *Biochim. biophys. Acta*, **367**, 282–294.
249. Petersen, O.H. (1974), *J. Physiol.*, **239**, 647–656.
250. Petersen, O.H. (1976), *Experientia*, **32**, 471–472.
251. Petersen, O.H. and Pedersen, G.L. (1974), *J. Memb. Biol.*, **16**, 353–362.
252. Philippot, J.R., Cooper, A.G. and Wallach, D.F.H., (1975), *Biochim. biophys. Acta*, **406**, 161–166.

253. Phillips, S.K. and Cramer, W.A. (1973), *Biochemistry,* **12**, 1170–1176.
254. Pierce, N.F., Greenough, W.B. III, and Carpenter, C.C. (1971), *Bact. Rev.,* **35**, 1–13.
255. Pietras, R.J., Naujokaitis, P.J. and Szego, C.M. (1975), *J. Supramolec. Struct.,* **3**, 391–400.
256. Pinteric, L., Manery, J.F., Chaudry, I.H. and Madapallimattam, G. (1975), *Blood,* **45**, 709–724.
257. Pinto da Silva, P. (1972), *J. Cell Biol.,* **53**, 777–787.
258. Pinto da Silva, P., Douglas, S.D. and Branton, D. (1971), *Nature,* **232**, 194–196.
259. Plate, C.A., Suit, J.L., Jetten, A.M. and Luria, S.E. (1974), *J. biol. Chem.,* **249**, 6138–6143.
260. Poo, M.M. and Cone, R.A. (1974), *Nature,* **247**, 438–441.
261. Poste, G., Papahadjopoulos, D., Jacobson, K. and Vail, W.J. (1975), *Biochim. biophys. Acta,* **394**, 520–539.
262. Poste, G., Papahadjopoulos, D. and Nicolson, G.L. (1975), *Proc. natn. Acad. Sci., U.S.A.,* **72**, 4430–4434.
263. Postel-Vinay, M.C., Sonenberg, M. and Swislocki, N.I. (1974), *Biochim. biophys. Acta,* **332**, 156–165.
264. Putney, J.W. Jr. (1976), *J. Pharmac. exp. Ther.,* **198**, 375–384.
265. Ranke, M.B., Stanley, C.A., Tenore, A., Rodbard, D., Bongiovanni, A.M. and Parks, J.S. (1976), *Endocrinology,* **99**, 1033–1045.
266. Rash, J.E. and Ellisman, M.H. (1974), *J. Cell Biol.,* **63**, 567–586.
267. Rasmussen, H. (1970), *Science,* **170**, 404–412.
268. Rasmussen, H. (1975), *Cell Membranes,* Chap. 20, (Weissman, G. and Claiborne, R., eds.), H.P. Publish. Co., Inc., New York.
269. Rasmussen, H., Jensen, P., Lake, W., Friedmann, N. and Goodman, D.B.P. (1975), *Adv. Cyclic Nucleotide Res.,* **5**, 375–394.
270. Rasmussen. H., Jensen, P., Lake, W. and Goodman, D.B.P., (1976), *Clin. Endocrinol.,* **5**, suppl., 11s.
271. Rasmussen, H., Lake, W. and Allen, E.J. (1975), *Biochim. biophys. Acta,* **411**, 63–73.
272. Reed, K., Vandlen, R., Bode, J., Duguid, J. and Raftery, M.A. (1975), *Archs. Biochem. Biophys.,* **167**, 138–144.
273. Reeves, P. (1965), *Aust. J. exptl. Biol. med. Sci.,* **43**, 191–220.
274. Reik, L., Petzold, G.L., Higgins, J.A., Greengard, P. and Barnett, R.J. (1970), *Science,* **168**, 382–384.
275. Rendell, M., Salomon, Y., Lin, M.C., Rodbell, M. and Berman, M. (1975), *J. biol. Chem.,* **250**, 4253–4260.
276. Révész, T. and Greaves, M. (1975), *Nature,* **257**, 103–106.
277. Révész, T. and Greaves, M.F. (1975), *Membrane Receptors of Lymphocytes,* (Seligmann, M., Preud'homme, J.L. and Kourilsky, I., eds.), North Holland Publish., Amsterdam, pp. 403–414.

278. Reynolds, J.A. (1972), *Ann. N.Y. Acad. Sci.*, **195**, 75–85.
279. Riggs, A.D., Bourgeois, S. and Cohn, M. (1970), *J. molec. Biol.*, **53**, 401–417.
280. Robison, G.A., Butcher, R.W. and Sutherland, E.W. (1967), *Ann. N.Y. Acad. Sci.*, **139**, 703–723.
281. Robison, G.A., Butcher, R.W. and Sutherland, E.W. (1971), *Cyclic AMP*, Academic Press, New York and London, p. 169.
282. Rogues, M., Tirard, A. and Lissitzky, S. (1975), *Molec. Cell Endocrinol.*, **2**, 303–316.
283. Romero, P.J. and Whittam, R. (1971), *J. Physiol.*, **214**, 481–507.
284. Root, A.W. and Oski, F.A. (1969), *J. Gerontol.*, **24**, 97–104.
285. Rosenheck, K., Lindner, P. and Pecht, I. (1975), *J. Memb. Biol.*, **20**, 1–12.
286. Rothman, J.E. and Dawidowicz, E.A. (1975), *Biochemistry*, **14**, 2809–2816.
287. Roy, C., Barth, T. and Jard, S. (1975), *J. biol. Chem.*, **250**, 3149–3156.
288. Rubin, M.S., Swislocki, N.I. and Sonenberg, M. (1973), *Archs. Biochem. Biophys.*, **157**, 252–259.
289. Sato, T., Garcia-Bunuel, R. and Brandes, D. (1974), *Lab. Invest.*, **30**, 222–229.
290. Scandella, C.J., Devaux, P. and McConnell, H.M. (1972), *Proc. natn. Acad. Sci., U.S.A.*, **69**, 2056–2060.
291. Schneider, A.S. (1973), *Meth. Enzymol.*, **27D**, 751–767.
292. Schneider, A.S. and Harmatz, D. (1976), *Biochemistry*, **15**, 4158–4162.
293. Schneider, M-J.T. and Schneider, A.S. (1972), *J. Memb. Biol.*, **9**, 127–140.
294. Schramm, M., Feinstein, H., Naim, E., Long, M. and Lasser, M. (1972), *Proc. natn. Acad. Sci., U.S.A.*, **69**, 523–527.
295. Schreier-Muccillo, S., Niculitcheff, G.X., Oliveira, M.M., Shimuta, S. and Paiva, A.C.M. (1974), *FEBS Letters*, **47**, 193–196.
296. Schwartz, I.L., Shlatz, L.J., Kinne-Saffran, E. and Kinne, R. (1974), *Proc. natn. Acad. Sci., U.S.A.*, **71**, 2595–2599.
297. Schwartz, S.A. and Helinski, D.R. (1971), *J. biol. Chem.*, **246**, 6318–6327.
298. Sedlacek, H.H., Stärk, J., Seiler, F.R., Ziegler, W. and Wiegandt, H. (1976), *FEBS Letters*, **61**, 272–276.
299. Sheppard, J.R. (1972), *Nature*, **236**, 14–16.
300. Sheridan, M.N. (1965), *J. Cell Biol.*, **24**, 129–141.
301. Shinitsky, M. and Inbar, M. (1976), *Biochim. biophys. Acta*, **433**, 133–149.
302. Shlatz, L. and Marinetti, G.V. (1972), *Biochim. biophys. Acta*, **290**, 70–83.
303. Shlatz, L. and Marinetti, G.V. (1972), *Science*, **176**, 175–177,
304. Shoemaker, W.C. and Elwyn, D.H. (1969), *Ann. Rev. Physiol.*, **31**, 227–268.
305. Siess, E.A., Nestorescu, M.L. and Wieland, O.H. (1974), *Diabetologia*, **10**, 439–477.
306. Singer, S.J. (1971), *Structure and Function of Biological Membranes*, Chap. 4, (Rothfield, L., ed.), Academic Press, New York.
307. Singer, S.J. (1975), *Cell Membranes*, Chap. 4, (Weismann, G. and Claiborne, R., eds.), H.P. Publishing Co., Inc., New York.

308. Singer, S.J. and Nicolson, G.L. (1972), *Science,* **175**, 720–731.
309. v. Smoluchowski, M. (1917), *Z. Physik. Chem.,* **92**, 129–168.
310. Sonenberg, M. (1969), *Biochem. biophys. Res. Commun.,* **36**, 450–455.
311. Sonenberg, M. (1971), *Proc. natn. Acad. Sci., U.S.A.,* **68**, 1051–1055.
312. Sonenberg, M. and Money, W.L. (1955), *Recent Prog. Hormone Res.,* **X1**, 43–82.
313. Sonenberg, M., Money, W.L., Dorans, J.F., Lucas, V. and Bourque, L. (1954), *Endocrinology,* **55**, 709–720.
314. Sonenberg, M., Swislocki, N.I., Bockman, R.S., Aizono, Y., Postel-Vinay, M.C., Rubin, M. and Roberts, J. (1973), *J. supramolec. Struct.,* **1**, 356–359.
315. Speth, V. and Wunderlich, F. (1973), *Biochim. biophys. Acta,* **291**, 621–628.
316. Stackpole, C. *Prog. Surface Sci.,* **12**, (in press).
317. Steck, T.L. (1974), *J. Cell Biol.,* **62**, 1–19.
318. Steck, T.L. and Fox, C.F. (1972), *Membrane Molecular Biology,* (Fox, C.F. and Keith, A., eds.), Sinauer Assoc., Inc., Stanford, Conn, pp. 32–35.
319. Stein, W.D. (1967), *The Movement of Molecules Across Cell Membranes,* Academic Press, New York.
320. Stier, A. and Sackmann, E. (1973), *Biochim. biophys. Acta,* **311**, 400–408.
321. Stoeckenius, W. and Engelman, D.M. (1969), *J. Cell Biol.,* **42**, 613–646.
322. Storm, D.R. and Chase, R.A. (1975), *J. biol. Chem.,* **250**, 2539–2545.
323. Storm, D.R. and Dolginow, Y.D. (1973), *J. biol. Chem.,* **248**, 5208–5210.
324. Tasaki, I. (1968), *Nerve Excitation,* C.C. Thomas Pub., Springfield, I11.
325. Taylor, R.B., Duffus, W.P.H., Raff, M.C. and DePetris, S. (1971), *Nature (New Biol.),* **233**, 225–229.
326. Taylor, S.I. (1975), *Biochemistry,* **14**, 2357–2361.
327. Tourtellotte, M.E., Branton, D. and Keith, A. (1970), *Proc. natn. Acad. Sci., U.S.A.,* **66**, 909–916.
328. Träuble, H. and Eibl, H. (1974), *Proc. natn. Acad. Sci., U.S.A.,* **71**, 214–219.
329. Träuble, H. and Sackmann, E. (1972), *J. Am. Chem. Soc.,* **94**, 4499–4510.
330. Triggle, D.J. (1971), *Neurotransmitter–Receptor Interaction,* Academic Press, New York.
331. Tsukamoto, T. and Sonenberg, M. (unpublished communication).
332. Tsushima, T., Shiu, R.P.C., Kelly, P.A. and Friesen, H.G. (1974), *D.H.E.W.,* (NIH) 74–612, 372–393.
333. Van Obberghen, E., DeMeyts, P. and Roth, J. (1976) *J. biol. Chem.,* **251**, 6844–6851.
334. Varga, J.M., Moellmann, G., Fritsch, P., Goodawska, E. and Lerner, A.B. (1976), *Proc. natn. Acad. Sci., U.S.A.,* **73**, 559–562.
335. Varga, J.M., Saper, M.A., Lerner, A.B. and Fritsch, P. (1976), *J. supramolec. Struct.,* **4**, 45–49.

336. Verkleij, A.J., Zwaal, R.F.A., Roelofsen, B., Comfurius, P., Kastelijn, D. and Van Deenen, L.L.M. (1973), *Biochim. biophys. Acta,* **323**, 178–193.
337. Vogel, Z. and Nirenberg, M. (1976), *Proc. natn. Acad. Sci., U.S.A.*, **73**, 1806–1810.
338. Volotovsky, I.D., Finin, V.S. and Konev, S.V. (1974), *Biofizika,* **19**, 666–669.
339. Wagner, R.C. and Bitensky, M. (1974), *Principles and Methods of Electron Microscopy of Enzymes,* (Hayat, M.A., ed.), Van Nostrand, New York pp. 110–131.
340. Wagner, R.C., Kreiner, P., Barrnett, R.J. and Bitensky, M.W. (1972), *Proc. natn. Acad. Sci., U.S.A.*, **69**, 3175–3179.
341. Wallach, D.F.H. and Zahler, P.H. (1966), *Proc. natn. Acad. Sci., U.S.A.*, **56**, 1552–1559.
342. Wendelaar Bonga, S.E. and Veenhuis, M. (1974), *J. Cell Sci.*, **16**, 687–701.
343. Wendt, L. (1970), *J. Bact.*, **104**, 1236–1241.
344. Williams, J.A. (1976), *J. Physiol.*, **260**, 105–115.
345. Williams, R.J.P. (1975), *Biological Membranes,* Chap. 7. (Parsons, D.S. ed.), Clarendon Press, Oxford.
346. Yahara, I. and Edelman, G.M. (1972), *Proc. natn. Acad. Sci., U.S.A.*, **69**, 608–612.
347. Yahara, I. and Edelman, G.M. (1973), *Exp. Cell Res.*, **81**, 143–155.
348. Yahara, I. and Edelman, G.M. (1973), *Nature,* **246**, 152–155.
349. Yahara, I. and Edelman, G.M. (1975), *Ann. N.Y. Acad. Sci.*, **253**, 455–469.
350. Yahara, I. and Edelman. G.M. (1975), *Exp. Cell Res.*, **91**, 125–142.
351. Yahara, I. and Edelman. G.M. (1975), *Proc. natn. Acad. Sci., U.S.A.*, **72**, 1579–1583.
352. Yamasaki, N., Kikutani, M. and Sonenberg, M. (1970), *Biochemistry,* **9**, 1107–1114.
353. Yin, H.H., Ukena, T.E. and Berlin, R.D. (1972), *Science,* **178**, 867–868.
354. Yoshida, S. and Ikegami, A. (1974), *Biochim. biophys. Acta,* **367**, 39–46.
355. Zipper, H. and Mawe, R.C. (1972), *Biochim. biophys. Acta,* **282**, 311–325.
356. Zwaal, R.F.A., Roelofsen, B. and Colley, C.M. (1973), *Biochim. biophys. Acta,* **300**, 159–182.
357. Zierler, K.L. (1972), *Handbook of Physiology and Endocrinology I,* Chapter 22, American Physiology Institute, Washington, D.C.

2

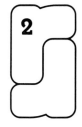

The Cellular Receptor for IgE

HENRY METZGER

National Institute of Arthritis,
Metabolism and Digestive Diseases,
National Institutes of Health,
Bethesda, Maryland

Receptors and Recognition, Series A, Volume 4
Edited by P. Cuatrecasas and M.F. Greaves
Published in 1977 by Chapman and Hall, 11 New Fetter Lane, London EC4P 4EE

2.1 INTRODUCTION

2.1.1 Description of system

Immunoglobulin E is a 4-chained immunoglobulin [7] which, like other immunoglobulins, is produced by plasma cells, but which is unique in its capacity to bind firmly to a surface 'receptor' on mast cells and related cells. The binding is not known to produce any perturbation of the cell until the cell is exposed to a substance (antigen or, in the laboratory, anti-IgE) which interacts with the IgE. Under appropriate conditions, this interaction leads to rapid partial or complete release of the cells' granule contents without concomitant cell death [5].

An electron micrograph of a sensitized (i.e. treated with IgE antibody) rat mast cell is illustrated in Fig. 2.1(a). A similar cell fixed 60 seconds after treatment with the appropriate antigen is shown in Fig. 2.1(b). Higher magnifications of these and similar preparations are presented in the informative article by Börje Uvnäs (96a).

The possible physiologic role of this process is unknowns but there is increasing evidence that it plays a significant patho-physiologic role in inflammatory reactions [5]. An apparent aberration of this function leads to allergic phenomena. IgE levels are also markedly increased in parasitic infections [51, 77] but the function of IgE under these circumstances remains obscure.

The mechanism by which the antigen (or anti-IgE)/IgE interaction on the surface of the cell leads to cell activation is uncertain. There is persuasive evidence that bridging or cross-linking of the surface IgE is required (Section 2.10.1) but how this generates an appropriate signal is unknown. A variety of biochemical steps have been implicated and their sequence assessed [52] but mostly by indirect (e.g. inhibition) studies. Ca^{2+} ions are clearly required (likely at two or more stages) and the reaction can be initiated by the injection of Ca^{2+} mechanically [53] or via ionophores [14].

This review concerns itself with what is known about the cell surface component which binds IgE. A considerable amount of information is now available about the IgE binding properties, structure, and membrane-integration of this component, but there is still no information about how it may be involved in the mechanism of degranulation. Several alternatives are considered at the end of this review.

2.1.2 General interest of the system

Several aspects of the system give it a wider interest. It is one of the more readily available systems in which the process of secretion can be readily studied [29]. It is at present perhaps the 'cleanest' system for studying an antigen-induced antibody-mediated cellular reaction: no other macromolecular or cellular components are implicated in the antigen-mediated triggering, the cell activation occurs rapidly and it can be measured quantitively and efficiently [88]. The wider applicability of facts learned from this system to the understanding other systems such as B and T cell triggering is presently rather an article of faith than fact.

2.2 MATERIALS AND METHODS

2.2.1 Cells

Direct binding studies have so far only been reported with rodent mast cells and related tumor cells. Initial success was reported with a unique rat basophilic leukemia (RBL-1) which arose in a Wistar strain rat fed the carcinogen β-chloroethylamine [30]. These cells have been successfully cultured [57] and by now are being used by a substantial number of investigators. Good preparations of cells from the minced tumors can be prepared by layering the cells on Hypaque-Ficoll and centrifuging briefly [16].* Three relatively recently induced mouse mastocytomas arising in pristane injected mice bearing an Abelson virus infection have been found to have good IgE binding activity [67]. One of these lines has been successfully cultured [67]. Two older mastocytomas have little or essentially no binding activity [57, 74]. Normal rat peritoneal mast cells are easily isolated and purified [93] the yield being $1-3 \times 10^6$ cells/rat. A 100-fold enrichment of mast cells from normal mouse peritoneal washings has been reported yielding preparations which were 65–75% toludine blue positive [95] but we have not had similar success [67]. Substantial purification of human basophils has also been claimed [22] but again has not been successfully reproduced (Lichtenstein, personal communication).

* The RBL-1 cells though containing histamine and granules have not been successfully 'triggered' to degranulate [90]. Nevertheless by presently available criteria their cell surface receptors for IgE are normal.

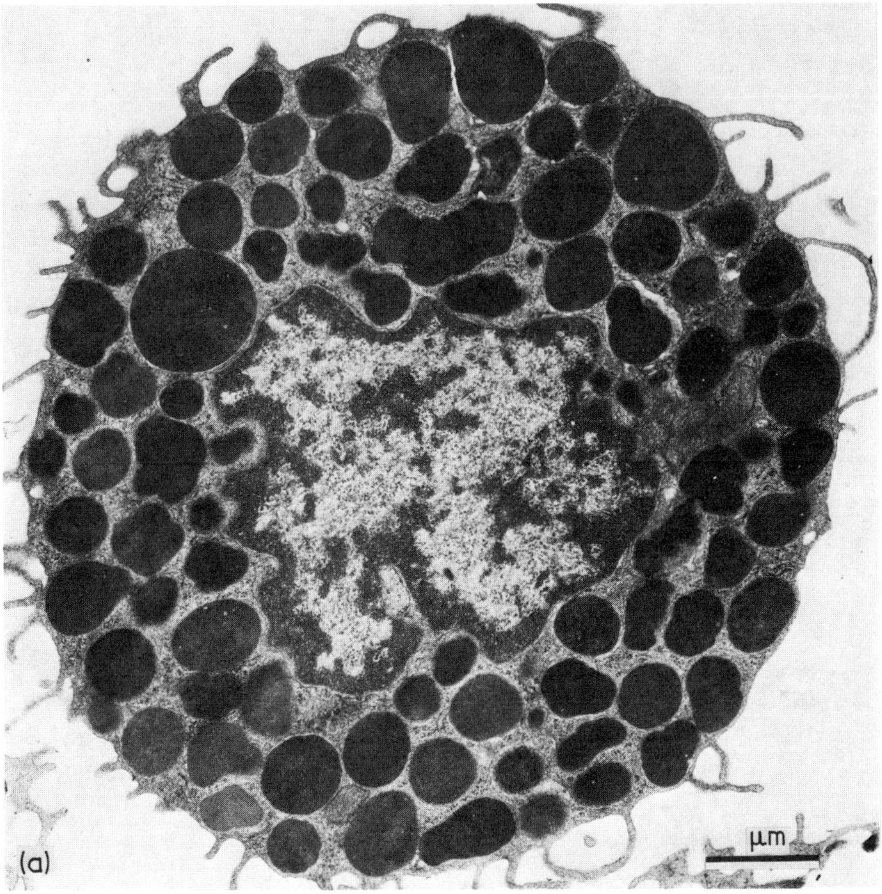

Fig. 2.1(a) Electron micrograph of a sensitized rat mast cell. Many homogeneous electron-dense (osmiophilic) granules are seen in the cytoplasm.

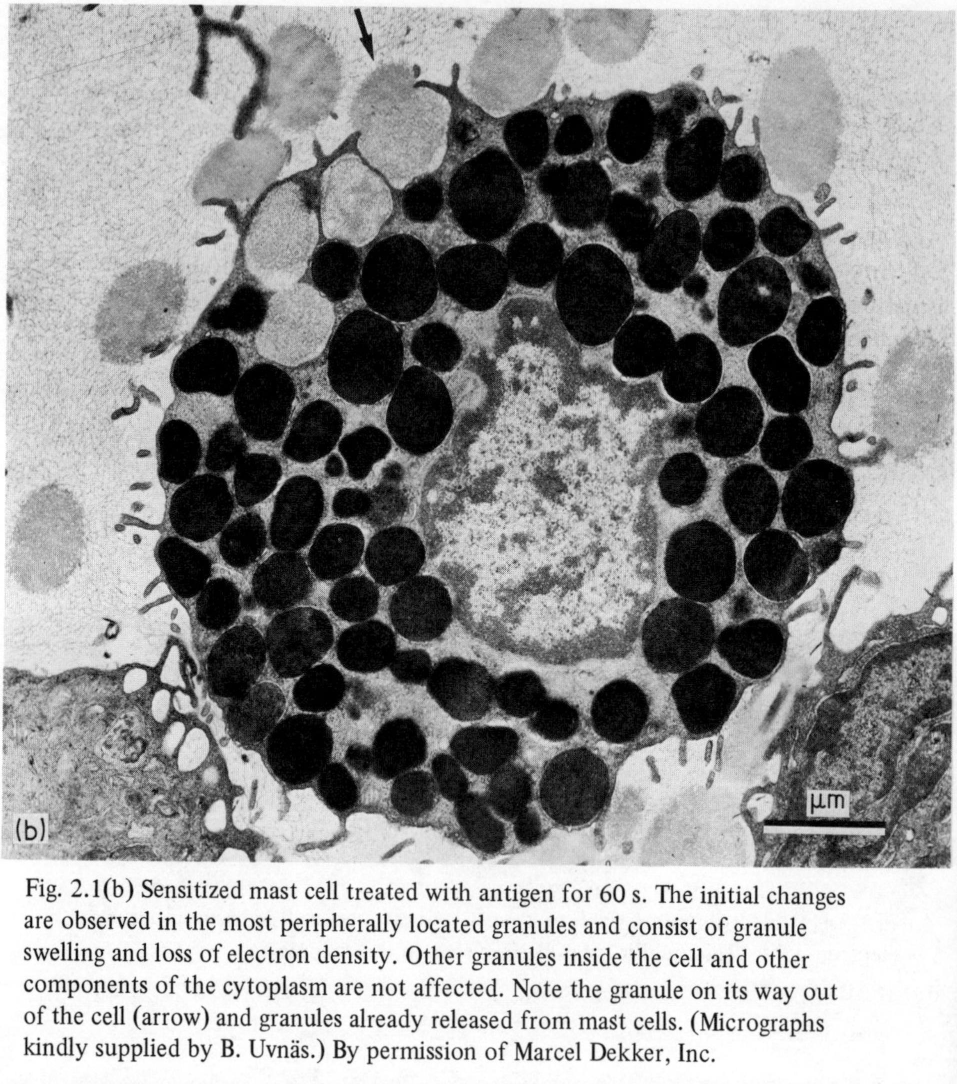

Fig. 2.1(b) Sensitized mast cell treated with antigen for 60 s. The initial changes are observed in the most peripherally located granules and consist of granule swelling and loss of electron density. Other granules inside the cell and other components of the cytoplasm are not affected. Note the granule on its way out of the cell (arrow) and granules already released from mast cells. (Micrographs kindly supplied by B. Uvnäs.) By permission of Marcel Dekker, Inc.

2.2.2 IgE

Normal rat IgE can be substantially purified by successive $(NH_4)_2SO_4$ precipitation, gel filtration and ion exchange chromatography using serum from parasite-infected rats [41]. Much larger amounts can be more easily prepared and more highly purified from the serum or ascitic fluid of rats bearing an IgE-producing tumor [4]. The IgE has a molecular weight of $\simeq 2 \times 10^5$ [4, 76] and that from tumor IR162 has an extinction coefficient (0.1%, 280 nm) of 1.36 [56]. Mouse IgE has as yet not been similarly characterized. There are no known IgE-producing mouse myeloma tumors. Several human IgE-producing myelomas have occurred and the proteins have been substantially characterized [7]. For optimal quantitative studies the IgE should have an accurately determined specific activity, show high bindability ($>$ 80%), and low non-specific binding ($<$ 5%) [56].

A limited number of studies have been performed with unpurified mouse reaginic antibodies. These studies suggest that their binding characteristics are similar to those of rat IgE [67].

2.2.3 Binding studies

Direct binding studies have been reported for intact cells, cell-free particulate preparations and detergent-solubilized receptors.

Cells

Well-washed cells can be incubated with radio-labeled IgE in the presence of a large excess of irrelevant protein. The cells can then be washed free of unbound IgE without affecting the bound molecules because of the very slow dissociation (Section 2.5.2). This is most simply done by sedimenting the cells through a fetal calf serum or other suitable 'cushion' [57, 66].

Membranes

IgE bound to membrane particles elutes at a very different position than unbound IgE when subjected to gel filtration [12, 55]. A simpler assay and one applicable to large numbers of samples uses microporous filters [71]. Such filters if adequately pre-treated with high concentrations of irrelevant proteins, fail to absorb free IgE but efficiently retain IgE bound to particulates.

Soluble receptors

IgE bound to detergent-solubilized receptors is insoluble at concentrations of $(NH_4)_2 SO_4$ at which unbound IgE is almost fully soluble. This permits a simple, rapid and accurate estimation of active soluble receptor concentration [84]. Some variation in relative precipitability of free and receptor-bound IgE is observed when the detergent concentration is varied. To increase the comparability of results from different experiments a 'positive control' in the form of known amounts of IgE-receptor complexes obtained by gel filtration (Section 2.7.2) can be incorporated in the proceduce [84].

2.3 DISTRIBUTION AND LOCATION OF RECEPTORS

Active receptors are diffusely distributed over the surface membrane of basophils and mast cells and not elsewhere. Studies utilizing labeled IgE and/or labeled anti-IgE antibodies demonstrated surface binding using fluorescence microscopy [63], electronmicroscopy [59, 91] and radioautography [46]. Using these techniques no evidence for local aggregates of receptors was found. Further evidence for surface binding comes from the observation that upon a brief exposure of the cells to acid pH the IgE is completely dissociated [48, 56]. Upon neutralization both the IgE and the cell-bound receptors are fully active [56, 70].

The fact that all of the active receptors are available on the surface was demonstrated by assessing the total number of sites on intact cells and comparing this to the number in a detergent extract of the same cells [84]. The number proved to be identical showing that there are no 'cryptic' active receptors elsewhere in the cells.

In microscopic studies, the IgE binding appears on basophils and/or mast cells exclusively [92]. IgE–antigen complex binding to eosinophils has been reported [34]. Although normal lymphocytes do not bind IgE there is a provocative report that receptors for IgE can be found on certain lymphoid lines [35]. Macrophages appear to have a receptor which binds aggregated IgE [9].

2.4 NUMBER OF RECEPTORS

Normal mast cells bind $\simeq 500\,000 \pm 300\,000$ molecules of IgE when the cells, isolated from normal rats or mice, are incubated with radioactive

IgE [15, 50, 67]. Cells from 'normal' animals prestripped of endogenous IgE by exposure to acid, show an additional $\simeq$ 10% binding [67]. Cells from parasite-infected animals show an expected reduction in the number of unsaturated sites due to their endogenous exposure to the higher serum IgE levels [50]. When purified mast cells are examined microscopically they show a remarkable heterogeneity in size and staining characteristics [18]. It would not be surprising if a co-ordinate variation in the numbers of IgE receptors were found but such an analysis has not been reported.

Such a study has been performed on the RBL-1 line of cells and even in this presumably homogeneous line marked variations were observed [42]. These changes correlated with the stage of the culture and with the cell-cycle distribution of the cells. Study of both synchronized and unsynchronized cultures suggested that the surface receptor became available for binding IgE during the G_1 phase of the cell cycle [50]. A substantial loss of binding activity was observed during the lag-phase of culture growth but the nature of this loss has not yet been ascertained.

An indirect method (Cl-transfer test [82]), which basically depends upon a stoichiometric 1:1 relationship between bound IgE, anti-IgE and complement components, was utilized to quantitate the IgE-binding capacity of human basophils from thirteen individuals [47]. Considerable variability was observed with the average being about one order of magnitude less than we have routinely observed with rodent mast cells. It would be worthwhile using the latter to compare the direct measurements with the indirect ones in order to determine if methodological factors account for the difference in numbers.

2.5 RECEPTOR–IgE INTERACTION

2.5.1 Mechanism

The binding of IgE to the receptor is via a simple bimolecular forward reaction and a first order dissociation [56]. This was demonstrated on intact cells by showing that the rate of binding was linear with regard to both the IgE and receptor concentration. Furthermore the binding curve was consistent with the theoretical curve constructed from the estimated rate constants and using the actual receptor and IgE concentrations. Similarly when cells which had been preincubated with radioactive IgE were incubated with a large excess of unlabeled IgE the loss of bound radioactivity was first order. With most cells studied, under conditions and over a time period where cell viability was well maintained, this loss amounted to only 10–20% because of the slow dissociation. However, normal mouse mast cells show a more rapid dissociation and 80–90% of the dissociation was observable [67]. In this case, also, the loss was first order. The reversibility of the binding is further demonstrated by the rapid dissociation of IgE from the cells after brief exposure to acid pH. As already noted, under suitable conditions both reactants are undamaged by this procedure.

2.5.2 Rate constants

Forward

The forward rate constant can be assessed by measuring the number of receptor-bound IgE molecules at early time points and utilizing the equation

$$k_1 = V_0/[\mathrm{IgE}]_0\ [\mathrm{Rec}]_0$$

Alternatively, more complete binding data may be fitted to the equation for the rate of a bimolecular reversible reaction [23]. The most detailed study of this type has been performed using RBL cells and the values ranged from $\simeq 1 \times 10^5\ M^{-1}s^{-1}$ estimated at 37° to $\simeq 2 \times 10^4\ M^{-1}s^{-1}$ at 4°C [56].

The interpretation of this value is not as simple as it may at first appear because the above equations are strictly speaking only applicable for reactions in which the reactants are homogeneously dispersed throughout the reaction volume. This is clearly not the case when studying intact cells. For example assuming 10^6 receptors/cell and 10^6 cells/ml the apparent receptor concentration is 10^{-9} M. However the receptors are actually localized in a total membrane volume of only about 4×10^{-6} ml (assuming a smooth sphere). Even given a highly villous surface [8] this would only add a factor of 2–4. Thus the receptor rather than being homogeneously distributed at 10^{-9} is distributed in 'packets' at $\simeq 10^{-4}$ M. The question arises to what extent the inhomogeneity of the receptor distribution affects the assumption that the calculated forward rate constant accurately reflects the intrinsic rate constant for the isolated receptor. A theoretical discussion of this problem is presented elsewhere [24]. Basically, we concluded that providing the time-dependent step was related to the intrinsic rate constant (as opposed to the rate of IgE diffusion) then there should be no substantial disparity. The availability of soluble receptors has now permitted an experimental determination.

A value of $\simeq 4 \times 10^5$ $M^{-1} s^{-1}$ at 4°C was estimated i.e. about an order of magnitude greater than found with the intact cells [84]. Additional studies would be necessary to clarify the basis for this discrepancy.

Reverse

When the reverse rate constant is estimated by incubating cells pre-incubated with radioactive IgE in the presence of unlabeled IgE, one cannot distinguish between the dissociation of IgE from receptor and dissociation of the receptor-IgE complex from the cell. The latter might be occurring because of cell shedding due to unfavorable growth conditions or normal turnover. In the former case the control cells (to which no unlabeled IgE had been added) would show a declining level of bound radioactivity while in the latter case such a drop would not be observed. Studies on the RBL cells suggest both phenomena may take place. Shedding was directly demonstrated under unfavorable growth conditions [12]. However even under conditions where cell viability was apparently well-maintained substantial differences in the 1/2-times of dissociation have been observed with intact cells (13–60 hrs) and even these values are considerably lower than those observed with membrane particles [72] or solubilized receptors [84]. So far it has only been possible to study the latter reliably at 4°C and a value of $\simeq 3 \times 10^{-7} s^{-1}$ was estimated. Thus calculating from k_1/k_{-1}, K_A for the detergent solubilized receptor is $> 10^{12}$ M^{-1} [84].

While more indirect studies indicated fairly substantial effects of variations in pH and a variety of salts, no major changes were seen when RBL-1 cells were studied similarly [56]. In any case small variations from physiological conditions are unlikely to affect the rate or extent of binding significantly.

A disparity in the reverse rate constant between normal mouse mast cells and mouse mastocytoma cells was documented by Mendoza and Metzger [67]. Although purified normal cells were not studied a variety of controls appeared to rule out binding to non-mast cells or other artifacts which could explain the more rapid dissociation from the normal cells. These findings raise the possibility that heterogeneity of the receptors with regard to their affinity for IgE may exist.

2.5.3 Association constant

Direct determination of binding constants has been reported for rat mast and RBL-1 cells using direct methods [50] and human basophils using indirect methods [47]. These values tend to be one or two orders of magnitude lower than those we have estimated from k_1/k_{-1}. We have elsewhere expressed our doubts about the accurracy of these lower values [67] particularly since they seem to require a k_{-1} which is clearly much higher than observed. One can only surmise that either equilibrium was not in fact achieved (it was assumed in the study on human cells) or receptor shedding or turnover was occuring.

2.5.4 Specificity

The receptor for IgE is highly specific. The binding is not affected by 1000–10 000 fold excess of other monomeric rodent immunoglobulins or of human IgE [38]. Conflicting reports on the ability of human IgE to bind to rat cells have appeared [79, 98]. At best the binding must be relatively weak and difficult to demonstrate reproducibly. A variety of experiments suggested that a subclass of rat IgG was able to bind to the same mast cell receptors [2,3]. A recent semi-quantitive study has shown that IgG can inhibit IgE binding only if the IgG is highly aggregated [38]. However under *in vivo* conditions even very small amounts of monomer IgG binding may be adequate to sensitize cells.

A recent report [61] challenges the previous data which suggested that IgG and IgE were competing for the same receptor. These workers found that the two hour passive cutaneous anaphylaxis reaction produced by skin sensitization with the IgG fraction of mouse serum could not be inhibited by an amount of rat myeloma IgE which was capable of inhibiting sensitization by IgE antibodies. An alternative explantation is as follows (suggested by R. Orange): the mast cell has a receptor which can bind either IgG (weakly) or IgE (strongly) and therefore inhibition of IgE binding by IgG is observed. Neutrophils also contain an Fc receptor for IgG but this does not bind IgE [45a 46]. Injection of skin with IgG would lead to sensitization of both types of cells but a positive reaction might be only partially inhibited by IgE. Thus the reaction Lehrer and collegues were observing might have been substantially due to release of slow-reacting substance of anaphylaxis (SRS-A) [5] rather than histamine release from mast cells. Orange's suggestion is now being tested.

2.6 INTEGRATION OF RECEPTOR IN MEMBRANE

Our knowledge of how the receptor is integrated into the membrane is still very limited (Metzger *et al.*, 1976). In terms of attachment the receptor has the attributed of an integral membrane protein: it is not dislodged by reagents which disrupt polar interactions but requires reagents which disrupt apolar bonds.

2.6.1 Exposure

Such integral proteins may be of two types: one in which a relatively small portion of the total protein is embedded in the membrane while the major portion is exposed to the solvent. An example of the latter type are the antigens coded for by the major histocompatability complex. Proteins of the second class are largely embedded in the membrane with only a relatively small portion exposed. The receptor for IgE behaves like a protein of this second type. Firstly, it is markedly resistant to proteolytic attack while on the membrane [71], although quite sensitive once solubilized [17, 58]. Less direct is the observation that while the free receptor is readily iodinated *in situ*, the IgE-complexed cell-bound receptor is not [17]. This suggests limited exposure of receptor tyrosines either because there are very few of them or because there is limited exposure of the receptor *per se.* Finally, observations on the antigenic determinants on the free receptor and receptor-IgE complex below and above the critical micelle concentration of detergent suggest the latter can almost totally mask most of the antigenic determinants (Section 2.8) [43]. To the extent that the reaction with detergent mimics the interaction with the membrane bilayer, these results also suggest a substantially embedded receptor.

2.6.2 Receptor–receptor interaction and valence

Ishizaka and Ishizaka [49] reported experiments with human basophils based on the apparent co-capping of IgE complexes and empty receptors, which suggested that each receptor might have more than one IgE binding site or that the receptors were in some way inter-connected. However the procedure was somewhat complex and a more straight-forward analysis of rodent mast cells failed to detect evidence of receptor multivalency or interaction. These experiments [68] involved saturating the cellular receptor with a mixture of fluorescein-conjugated and rhodamine-conjugated IgE. When the cells were reacted with anti-IgE the fluorochromes were co-patched or co-capped. However, when reacted with anti-fluorescein only the fluorescein-conjugated IgE was redistributed and the rhodamine-conjugated IgE remained diffusely distributed over the cell surface. Similar findings were observed with normal and tumor rat and mouse cells. These findings seem to demonstrate unambiguously that the receptor is monovalent with respect to IgE and at least prior to aggregation is unconnected to other receptors (see also Section 2.7.3). Using a similar approach, these observations were confirmed using the method of photo-bleaching of fluorescence (Section 2.6.3) [85].

2.6.3 Mobility and interaction with other cell components

Antibody-induced redistribution

The experiment described above demonstrated that, as with other cell surface determinants and other cells, cross-linking can induce gross redistribution of the surface component. The IgE receptor system is of interest in this regard because one can systematically vary the number of surface components (by varying the receptor occupancy with IgE) and the cross-linking reagent (anti-IgE) and study how this affects both aggregation and redistribution [11]. This is of particular interest in relation to IgE-mediated cell triggering, and the system would seem ripe for such a systematic investigation (see also Section 2.10). It is also of interest that the pattern of redistribution appears to be sensitive to the number of cross-links per surface component (IgE). Thus it was observed that when cells were saturated with fluoresceinated IgE and reacted with anti-IgE, large patches or caps were formed on normal mast cells or the tumor cells respectively. However, when the identical preparations were reacted with anti-fluorescein the pattern was quite different–the patches were much smaller with both types of cells and apparent endocytosis was prominent. The patterns were not interconvertable over a range of antibody doses [68].

Photo-bleaching of fluorescence

In addition to studying the movement of the IgE–receptor complex induced by antibodies, it is now possible to study the movements of the unperturbed complexes. The method utilized, photo-bleaching of fluorescence, involves labeling surface components with a fluorochrome, focal irradiation of a single cell with a high intensity laser beam (thereby irreversibly bleaching the fluorochromes in the irradiated area) and then observing the recovery of fluorescence over the same area [32, 85]. Recovery occurs if the bleached molecules are free to move out of the area and unbleached molecules are free to enter. From the kinetics and equilibrium value of fluorescence recovery as well as the effect of various drugs on these values one may gain insight into the relation of the cell surface component to other membrane or submembrane constituents. Such experiments had been carried out with non-specific fluorescein-conjugated cells or with rhodamine-labeled concanavalin A [32, 85]. The studies with rhodamine-labeled IgE on normal mast cells [86] represents one of the first investigations in which one was able to examine a discrete, unaggregated cell surface component.

The initial studies suggested that the receptor motility was heterogeneous, partially restricted and markedly inhibitable by cytochalasin B. The data must be extended and supplemented by other approaches but these initial observations would certainly be consistent with receptor – microfilament interactions.

The studies utilizing both the photo-bleaching of fluorescence method and the antibody-induced redistribution are informative with regard to the IgE–receptor complex. Neither can give information, however, on whether the IgE binding itself leads to changes in the interactions between the receptor and other cell components. *A priori* there is no reason to suspect that such changes occur but this point should be investigated.

2.7 STRUCTURE OF SOLUBILIZED RECEPTOR

2.7.1 Composition

There are as yet no published reports of adequate purification of the receptor and no compositional analysis is yet possible. Nevertheless, indirect data suggest that it is a glycoprotein: (a) it is exposed on the surface membrane, (b) it is iodinatable [17], (c) radioactive amino acid and carbohydrate residues are incorporated [58], (d) it is sensitive to proteases [17, 58], (e) it shows a steep temperature and pH dependence of denaturation [71], (f) its activity and gel filtration behaviour are unaffected by treatment with a 'cocktail' of lipases (G. Rossi, unpublished observations).

2.7.2 Size

Native receptor

If a variety of proteins of known molecular weight are chromatographed on a suitable gel and the elution volume of each plotted *versus* the log of the molecular weight a straight line results. With serum or other 'soluble' proteins no significant change is observed in the relative position of the proteins if the experiment is run in the presence of 1% Nonidet P-40 or similar detergents [13]. If a solubilized receptor preparation is run under similar conditions and the effluent monitored for IgE binding activity a discrete symmetrical and narrow peak of activity is observed [84]. From this type of analysis performed on Sepharose 6B at 4°C an apparent molecular weight of 250 000 can be calculated. If IgE–receptor complexes are similarly analyzed a molecular weight of 410 000 can be estimated. Using a similar approach Conrad and Froese estimated the molecular weight of the complex as between 350 000 and 550 000 [17]. The difference between the free and complexed receptor – about 160 000 – is roughly comparable to the molecular weight of IgE estimated by the same technique i.e. about 200 000. These results first suggested that the solubilized receptor was univalent [84].

Such an analysis has serious deficiencies, however, because the elution volume is strictly speaking related to the Stokes radius of a particle and only directly related to the mass by the degree to which the particle assumes a spherical shape.

If, instead, the column data are plotted *versus* the Stokes radius (or an equivalent function $1/D$ where D is the diffusion constant) the data are again linear and a diffusion constant for the free receptor and IgE–receptor complexed can be estimated (Table 2.1). This may then be used in conjunction with the Svedberg equation to calculate a molecular weight providing the sedimentation rate and partial specific volume ($\bar{v}$) can be determined [13].

The latter term provides the greatest difficulty since a direct measurement of $\bar{v}$ is impossible when only microgram or submicrogram amounts of the component are available. An alternative approach is to determine the sedimentation rate in H_2O and D_2O and using the shift in the rate of migration to estimate the buoyancy factor i.e. $\bar{v}$ [31, 73]. The estimation must be very accurate for the following reason. If a substantial amount of detergent is bound by the particle, the $\bar{v}$ of the receptor–detergent complex (which is what is actually measured of course) is very high since the $\bar{v}$ for non-ionic detergents is close to unity. Thus,

Table 2.1 Molecular parameters of native receptor and receptor–IgE complex in non-ionic detergent*

	$D_{20,w}$ ($\times 10^7$)	$\bar{v}$	$S_{20,w}$	Mol. wt.
Receptor	3.2	0.81	3.3	130 000
Receptor–IgE	2.5	0.78	7.3	310 000
IgE	3.3	0.74	7.0	200 000

*Adapted from [76].

since in the determination of the sedimentation rate and the molecular weight the term $(1-\bar{v}\rho)$ is applied, a 1% error in $\bar{v}$ leads to a $\simeq$ 5% error in the calculated molecular weight. It is beyond the scope of this review to discuss these points in greater detail here. A useful analysis of these problems has been presented by Clarke [13] and their application to the receptor for IgE is given in our own study [76]. Table 2.1 gives our best estimates of the molecular weight of the receptor–detergent, and the receptor–detergent–IgE complex using these techniques. The results are clearly consistent with univalency of the receptor, and with a molecular weight of the detergent–receptor of about 130 000.

If larger amounts of receptor were available one could estimate the detergent contribution to the molecular weight directly. Since such quantities are not available it can only be guessed at by *assuming* a $\bar{v}$ for the detergent-free receptor. Ultimately, compositional data are necessary to determine the validity of this assumption. Using a value that is typical for serum glycoproteins we estimate a molecular weight of 77 000 for the detergent-free receptor (see also Section 2.7.2). One further reservation should be mentioned. The calculations assume that the receptor and receptor–IgE complex bind equivalent quantities of detergent. Since the IgE does not appear to bind significant detergent this is a reasonable assumption.

Denatured receptor

Several membrane proteins have been usefully analyzed as follows. The protein on (in) the intact cell is labeled (along with other proteins) either by surface labeling e.g. lactoperoxidase-catalyzed iodination, or by incorporation of radioactive amino acids or sugars. The cell is then disrupted e.g. with detergents, and the desired component specifically precipitated with antibody. Usually double immunoprecipitation is used.

Table 2.2 Analysis of denatured receptor by polyacrylamide gel electrophoresis

% cross-linked	Apparent mol. wt.	Reference
5	62 000	[1]
5.9	60 000–70 000	[58]
7.5	57 000 ± 600	*
10.0	54 000 ± 300	*
10–12	⩽52 000	[58]
12.5	53 000 ± 200	*
15	52 000 ± 500	*

* Taurog and Metzger, unpublished observations.

The precipitates are then dissolved in sodium dodecyl sulfate and the solution analyzed by polyacrylamide gel electrophoresis. A similar approach to the IgE receptor was first employed by Conrad and Froese [16]. Rat mast cells or RBL cells were surface-iodinated, saturated with IgE, and solubilized. The extract was then reacted with either rabbit anti-IgE or normal rabbit IgG and then with goat anti-rabbit IgG. The resulting precipitates were analyzed on 5% gels and showed primarily a single band with an apparent molecular weight of 62 000. Kulczycki *et al.*, [58] obtained similar results. At higher acrylamide concentrations the estimated molecular weight is about 10 000 lower (Table 2.2). Similar results are obtained if the receptor is internally labeled with radioactive amino acids or sugars [58]. This suggests that the surface labeling method is also labeling a more deeply embedded component [16].

The gel electrophoresis data are not without their own ambiguities. Several striking errors have been made with this method. For example, J-chain which has a molecular weight of about 15 000 was originally estimated at 25 000 (error = 10 000 daltons, + 67%) and the major red cell ghost glycoprotein of molecular weight 29 000 was estimated as 55 000 by acrylamide gel electrophoresis (error = 26 000 daltons, + 90%). The problem is that if the bound detergent/unit weight is not the same in the unknown as compared with the standards (1.4 $g\ g^{-1}$) the migration rate will be aberrant. Glycoproteins appear to bind less detergent. Therefore at low degrees of gel cross-linking where the migration may be more sensitive to the charge than to the size, the relative migration is lower leading to an over-estimate of the size. At higher degrees of cross-linking where the size limits the migration rate more severely the error should be smaller. It is likely that this accounts for the decrease in molecular weight estimated seen in Table 2.2. However, even the lower values might still be far too high.

Possible models

If one tries to rationalize the data on the native and denatured receptor the following models appear reasonable.

(1) If the receptor contains substantial carbohydrate the $\bar{v}$ might be as low as 0.68 (the value for the major glycoprotein of red cells [36]). In this case the mass of the detergent-free receptor would be about 63 000 daltons. In turn the gel data could be substantially aberrant and the true mass of the denatured polypeptide would be as low as 30 000 daltons. Thus the native receptor would be a dimer.

(2) If the receptor contained very little carbohydrate the $\bar{v}$ could be as high as 0.76, giving a calculated molecular weight of the native receptor of about 90 000. The gel data at high levels of cross-linking would then be approximately correct i.e. $\simeq$ 50 000. Again a dimeric structure would be most consistent.

(3) The receptor could be monomeric, the discrepancy simply being due to errors in the measurements. If this were so, the determination of the molecular weight of the native receptor would be most suspect since the measurements and calculations are considerably more complex.

Two approaches seem most reasonable to resolve this problem

(a) Isolation of the purified receptor and determination of its composition.

(b) Reaction of the purified receptor with cross-linking reagents. The latter approach has been shown to be useful for determining the subunit structure of several oligomeric proteins [10, 21].

2.7.3 Valence

This subject has already been touched upon in a different context. Despite the ambiguity in the molecular weight data, the solubilized receptors clearly seem to be univalent [76]. This is consistent with the apparent univalency of the receptor *in situ* [67]. It demonstrates that the isolated receptor is unaggregated.

2.7.4 Interaction with detergents

The experiments on the solubilized receptor discussed in Section 2.8.2 were all conducted at 4°C. The estimated detergent binding is approximately 0.7 mg mg^{-1} [76] not out of line with some other membrane proteins determined at the same temperature [13]. If the experiments are conducted at room temperature, a substantial apparent increase in molecular weight is seen (Newman, unpublished observations). Preliminary data suggest this is solely due to an increase in bound detergent. Whether this is due to exposure of new detergent-binding sites due to unfolding of the protein or related to the increase in micelle aggregate size with temperature (implying that it is the micelle rather than monomer which reacts with the receptor, at least at the higher temperatures) remains to be determined. In this context, it is of interest that the masking of antigenic determinants on the receptor becomes evident at approximately the critical micelle concentration and continues to become more profound at concentrations of detergent where the micelle but not the monomer concentration of detergent is increasing. This is also suggestive that it may be micelles which are being bound. These observations suggest that it may be invalid to calculate the hydrophobic surface area of a membrane protein simply on the basis of the amount of detergent bound [13] particularly if experiments at 4°C are used to interprete structures occuring at 37°C.

2.7.5 Other features of receptor structure

A variety of other features of the receptor may be added for sake of completeness. The binding activity of the receptor is quite unstable at temperatures above 4°C the half life being 10–15 minutes at 37°C in some preparations [84]. This is somewhat variable however, and this fact plus the observation that partially purified receptor is usually more stable suggests an extrinsic cause. Presumably this is a proteolytic enzyme although attempts to inactivate such an enzyme have been unsuccessful (Rossi, unpublished observations). It is somewhat surprising that the antigenic determinants on such inactive receptors are largely preserved [43].

It is of interest that when complexed to IgE the receptor is considerably more stable [84]. Whether this is due to shielding of protease-sensitive bonds, to a conformational change, or stabilization is unclear.

2.8 REACTION OF ANTIBODIES WITH RECEPTOR

An antiserum to mast cells was found to precipitate several radioactive proteins from a surface-labeled preparation among which was a component with the same electrophoretic mobility as the receptor component

[33, 99].

A similarly prepared antiserum was analyzed in an alternative manner [43]. It was shown to contain antibodies to the receptor by the capacity of derived Fab fragments to inhibit binding of IgE to the receptor *in situ* and to the non-ionic detergent-solubilized receptor. In addition the serum could totally precipitate the IgE binding activity of a solubilized cell preparation. The antiserum also reacted with receptor–IgE complexes a little less efficiently. These results suggest that the antiserum was reacting with at least two (classes of) determinants – one in or near the IgE binding site (from which it could be displaced by sufficient IgE) and another more distal. In the presence of detergent above the critical micelle concentration, the reactivity with unoccupied receptors became less efficient while with receptor-IgE complex it was more or less completely eliminated. Detergent had no effect on the reactivity of anti-IgE with the receptor–IgE complex. Taken together these results suggest that the second (class of) determinant was progressively masked by detergent. Although there are other possible explanations, one that would be consistent with other data is that this reflects the relatively large proportion of the receptor mass which is embedded in the membrane (Section 2.6.1).

There are a variety of useful experiments which could be performed with anti-receptor antibody if it can be made monospecific [33, 43].

2.9 IgE STRUCTURE IN RELATION TO RECEPTOR BINDING

2.9.1 Location of binding domain

More or less direct studies showed that the binding of human IgE to the mast cell receptor was due to the Fc region [91, 45]. A dimeric fragment which included the $C_{\epsilon 2}$ domains ($F[ab']_2$) was inactive as was the Fc′ fragment which corresponds roughly to the $C_{\epsilon 2}$ domain alone [7]. Similarly fragments corresponding to the $C_{\epsilon 4}$ region alone are said not to have cytotropic activity (cited in [27]). The irreversible inactivation of IgE with regard to its cytophilic properties by reduction or heating has long been known. Dorrington and Bennich [27] attempted to localize the site of irreversible changes as judged by cirular dichroism and difference spectra. They concluded that $C_{\epsilon 3}$ was the primary region where irreversible changes took place so that this area is implicated as being important. Unfortunately, it is very difficult to perform quantitative binding studies in the human system because adequately homogeneous cell populations are not available. Such binding data provide the most adequate assay for localizing the binding domain. In the rat, where one has excellent binding assays and now plentiful supplies of IgE it has so far not been possible to even obtain an active Fc fragment (Kanellopoulos, unpublished observations).

2.9.2 The pentapeptide story

Considerable interest was evoked by the report that claimed inhibition of IgE-induced skin sensitization (Prausnitz-Kustner reaction) by a pentapeptide whose sequence correspond to residues 320–324 in the $C_{\epsilon 2}$ domain of IgE [39]. The results were interpreted as being due to inhibition of IgE binding by the short peptide.

There are several reasons why this provocative interpretation cannot yet be accepted as proven.

(1) The inhibition of the skin reaction was only partial, variable and not always observed (Hamburger, personal communication).

(2) Surprisingly little specificity was observed when sequence analogues were tested.

(3) Extremely high molar ratios (up to 10^6-fold) of peptide to reagin were required to see palpable effects.

(4) The sequence comes from a region which by other techniques appears to be inactive. That is, the $F(ab')_2$ fragment which contains this sequence is inactive [7] though it is true the fragment was not tested at such high doses.

(5) An apparent confirmation of Hamburger's result [97] is highly suspect. The latter workers claimed that they could desensitize rat-IgE sensitized rat mast cells with uncertain doses of the human peptide after ten minutes incubation. The binding of human IgE to rat mast cells is very weak at best (Section 2.5.4) and since even high doses of free intact rat IgE cannot displace cell-bound IgE from rat mast cells in anywhere near this time period, it is simply not possible that the effect observed was due to competitive binding by the peptide. A 10^6-fold excess of the properly-neutralized peptide (obtained from R.N. Hamburger) failed to have any effect on the binding of rat IgE to RBL cells in our hands (Mendoza and Metzger, unpublished observations, 1975). A demonstration of the peptide's effectiveness in preventing a positive skin test by Stanworth at a workshop in Amsterdam, 1975, remains unpublished as does the claim of its complete ineffectiveness (H. Bennich, personal communication).

(6) The peptide's capacity to inhibit a skin reaction remains as the only evidence of its activity to date. A variety of other mechanisms e.g. pharmacological effects, have not been ruled out.

2.10 FUNCTIONAL ASPECTS

2.10.1 Requirement for cross-linking

In many cell-surface receptor systems it appears that either the receptor itself undergoes a conformational change which serves as the 'signal' or that the ligand is changed and by entering the cell serves as the stimulus.

In the IgE-mast cell system it appears that the signal is generated by a cross-linking of the surface IgE. The evidence for this is assembled in Table 2.3. Since cross-linking is required for 'activation' and is a prerequisite for surface redistribution [63, 94] it is logical to ask whether gross redistribution is required for activation. The evidence is persuasive that it is not [6, 59]. Indeed, if redistribution occurs within the time span that degranulation would ordinarily be expected, release of granule contents is in fact inhibited [26, 64, 65]. Taken together these results

Table 2.3 Evidence for bridging mechanism as signal for IgE-mediated exocytosis

Stimulus	Cell sensitized with	Exocytosis	Reference
Univalent x	Anti-x	–*	[64, 75, 89]
Multivalent x	Anti-x	+	[64, 75, 89]
x–y	Anti-x	–	[64, 75, 89]
x–y	Anti-y	–	[64, 75, 89]
x–y	Anti-x plus anti-y	+	[64, 75, 89]
Fab of anti-IgE	IgE	–	[6, 44–46, 59, 65]
IgG of anti-IgE	IgE	+	[6, 44–46, 59, 65]
Excess IgG of anti-IgE	IgE	– redistribution	[6, 65]
Monomeric Fc_{ϵ}	None	–	[45]
Aggregated Fc_{ϵ}	None	+	[45]

* Those experiments (reviewed in [80]) in which univalent ligands were apparently capable of initiating exocytosis are ambiguous since it is possible that the ligands were bound to the surface of other cells, thereby exposing the sensitized cells to effectively polyfunctional stimuli [25].

suggest that the distribution of the surface IgE in space and time are important with regard to the exocytotic phenomenon. The next section describes some alternatives for how aggregation of the surface IgE might generate a signal.

2.10.2 Alternative models

As noted in the introduction there are no data which directly or indirectly show that the component which binds IgE is itself part of the mechanism by which antigen (or anti-IgE) induces IgE-mediated cell degranulation. It may be useful to outline alternative roles that this component might play and to indicate experimental approaches by which one could decide which alternative is correct.

Model I: Anchoring role
By definition the component which binds IgE anchors the IgE to the surface membrane. It could be that that is all it does. Antigen or antibody-induced aggregation of the IgE resulting in the aggregation of Fc_ϵ could lead to some direct perturbation (e.g. formation of an ion channel) or enchanced interaction of a true receptor with the IgE directly – not via the IgE-binding component.

Model II: Linkage role
The aggregation of the IgE binding protein via IgE aggregation could alter the interaction between it and another component in or below the membrane. Perturbation of the latter could generate the signal.

Model III: Direct role
The aggregation of the IgE binding protein via IgE aggregation could be sufficient to produce an effect directly; an ion channel might form, or an intrinsic enzymatic activity be unmasked. Rosenstreich and Blumenthal [83] recently compared the ionophorous effect on lipid bilayers and the mitogenic effect on lymphocytes for various compounds. They observed that certain polymeric proteins (e.g. keyhole limpet hemocyanin, excitability inducing material [83]) were active but only in their polymeric state. These intriguing observations may be related to the speculative models described above.

2.10.3 Experimental approach

A variety of experimental explorations may be useful for elucidating the true function of the IgE-binding protein. The anticipated results from three such approaches are listed in Table 2.4. The first approach is simplest and is possible to perform rigorously as soon as monospecific anti-receptor antibodies become available.

Table 2.4 Predicted (+) or unpredicted (–) results related to role of IgE-binding cell surface protein

Experimental approach	Model		
	I	II	III
Cross-link with anti-receptor, assay for histamine release	–	+	+
Search for ion binding, enzymatic activity, etc. of purified aggregated receptor	–	–	+
Search for interaction of other cell component with:			
IgE	+	–	–
Receptor	–	+	±

2.11 COMPARISON WITH OTHER 'Fc' RECEPTORS

2.11.1 Cell distribution

Binding activity for aggregated IgG has been observed on a large number of cells: B-lymphocytes, activated T-lymphocytes, macrophages, neutrophils, intestinal epithelial cells and hepatocytes, and viral-infected fibroblasts [28, 40, 54, 69, 100]. As previously described, the receptor for IgE on mast cells and basophils is likely to be the same as the receptor for IgE. However except for the interesting observation of Gonzales-Molina and Spiegelberg, [35] the other receptors for IgG do not appear to bind monomeric IgE. It is beyond the scope of this review to comprehensively compare all of these. A review on lymphocytes has been published recently [26].

2.11.2 Binding characteristics

Almost all the receptors for non-IgE Fc bind monomeric IgG only weakly. An exception is the receptor on macrophages an example of which has recently been carefully examined by Unkeless and Eisen [96]. Their data are compared to the data on the rodent IgE receptors and to more sparce data on other Fc receptors in Table 2.5. The Fc_γ receptor on gut epithelial cells may be another example of an Fc receptor of high affinity for monomeric IgG [37].

Unlike the receptor for Fc_ϵ, Fc_γ receptors generally show limited or no species specificity. Variable results have been reported in binding studies of the IgG subclass specificity. The macrophage receptor studied by Unkeless and Eisen, on the other hand, clearly distinguishes between IgG_{2a} and IgG_{2b} and IgG_1. As with most other receptors for Fc_γ examined, IgA and IgM were not bound at all. (Separate receptors on monocytes for IgM have however been reported, [60]). In all cases where it has been studied the binding of Fc_γ is exothermic [96] as it is for Fc_ϵ [56].

2.11.3 Structural data

Rast and co-workers [81] treated crude membrane fractions from mouse spleens with EDTA and mercapto-ethanol, recovered the 100 000 g supernatant and applied this to a Sepharose column which had been coupled with aggregated IgG. The material eluted from the column was labeled with ^{125}I or ^{3}H-borohydride and subjected to polyacrylamide gel electrophoresis and gel filtration in 6 M guanidine–HC1. A component was observed with an apparent molecular weight of 65 000 as well as smaller components which were thought to be breakdown products. The material isolated from the Sepharose column was capable of reacting with aggregated IgG and antibodies to it blocked Fc receptor activity on intact cells. It is unclear whether the material isolated represents aggregated Fc receptors, monomer Fc receptors which have multiple combining sites or monomer Fc receptor which are univalent. The latter would be surprising in view of the low affinity of the Fc receptor for monomeric Fc.

Cooper and Sambray [19] studied the shed material from surface-labeled murine leukemia cells reacted with aggregated human IgG. The released counts were analyzed on (5%) polyacrylamide gel electrophoresis in sodium dodecyl sulfate and it was claimed that a component of molecular weight 45 000 could be isolated. Documentation for this

Table 5.5 Comparative binding data on cellular Fc receptors*

Cells			Ig studied		k_1	k_{-1}	K_A	Sites per cell	Reference
Line	Type	Species	Class	Species	$(M^{-1}s^{-1} \times 10^{-1})$	$(s^{-1} \times 10^{5})$	(M^{-1})	$(\times 10^{-5})$	
RBL	Basophilic leukemia	Rat	IgE	Rat	1	$\leqslant 1$	$\geqslant 10^{10}$	1–15	[56]
$P388D_1$	Macrophage	Mouse	IgG_{2a}	Mouse	1.3	430	3×10^7	~1	[96]
ARTB–1	Lymphomat †	Mouse	IgG	Rabbit	N.D. ‡	N.D.	$2–5 \times 10^5$ §	~1	[87]
–	Macrophage (Peritoneal exudate)	Guinea pig	IgG_2	Guinea pig	N.D.	N.D.	$\sim 1 \times 10^6$ ¶	25	[62]
–	Macrophage (Alveolar)	Rabbit	IgG	Rabbit	N.D.	N.D.	$\sim 1 \times 10^6$	20	[1, 78]

* All data for 37°C
† This tumor line has mixed T and B cell characteristics
‡ N.D.– No data
§ Estimated from 0°C data and temperature dependence
¶ Estimated from 20°C data and presumed temperature dependence

estimate is lacking.

The Fc receptor on macrophages is very sensitive to tryptic hydrolysis in some studies and not in others, where as lymphocyte Fc receptors are characteristically not. As with the receptor for Fc_ϵ such data can be provisionally interpreted as reflecting the degree to which a substantial portion of the receptor is embedded in the surface membrane.

2.11.4 Functional aspects

It is clear that the receptor for Fc_γ is critical in antibody-dependent cell cytotoxicity by lymphocytes, phagocytosis by macrophages, and receptor-mediated transport in certain epithelial cells. A variety of other biological roles can be envisioned. The mechanism of any of these remains obscure. Potentially one of the most interesting Fc receptors is the putative component which binds the endogenous surface immunoglobulin on B-cells. However, there are in fact no data as to whether it exists or not and certain theories of B-cell activation don't require that such a receptor be present [20].

2.12 CONCLUDING REMARKS

In this review, which was completed in October 1976, I have summarized our current state of knowledge about the surface receptor for IgE. Over the past several years considerable information has been obtained, though as indicated two very basic items – the receptors overall role in triggering and it's composition – remain undefined. It appears likely that these points will soon be resolved but one of the most interesting aspects, i.e. the problem of defining precisely the triggering mechanism poses a formidible challenge. One only needs to recall how difficult it has been to obtain such knowledge about even such a 'simple' system as O_2 binding to hemoglobulin – this despite the availability of grams of material, detailed X-ray structures, innumerable homologues to compare, sensitive assay systems which permit detailed kinetic analyses etc. It is difficult to estimate at this time when comparable information on the receptor for IgE will be obtained.

REFERENCES

1. Arend, W.P. and Mannik, M. (1973), *J. Immunol.*, **110**, 1455–1463.
2. Bach, M.K., Bloch, K.J. and Austen, K.F. (1971), *J. exp. Med.*, **133**, 752–771.
3. Bach, M.K., Bloch, K.J. and Austen, K.F. (1971), *J. exp. Med.*, **133**, 772–784.
4. Bazin, H., Querijean, P., Beckers, A., Heremans, J.F. and Dessy, F. (1974), *Immunol.*, **26**, 713–723.
5. Becker, E.L. and Henson, P.M. (1973), *Adv. Immunol.*, **17**, 94–193.
6. Becker, K.E., Ishizaka, T., Metzger, H., Ishizaka, K. and Grimley, P.M. (1973), *J. exp. Med.*, **138**, 394–409.
7. Bennich, H. and Johansson, S.G.O. (1971), *Adv. Immunol.*, **13**, 1–55.
8. Buell, D.N., Fowlkes, B.J., Metzger, H. and Isersky, C. (1976), *Cancer Res.*, **36**, 3131–3137.
9. Capron, A., Dessaint, J.P., Capron, M. and Bazin, H. (1975), *Nature*, **253**, 474–475.
10. Carpenter, F.H. and Harrington, K.T. (1972), *J. biol. Chem.*, **247**, 5580–5586.
11. Carson, D. and Metzger, H. (1974), *J. Immunol.*, **113**, 1271–1277.
12. Carson, D., Kulczycki, Jr., A. and Metzger, H. (1975), *J. Immunol.*, **114**, 158–160.
13. Clarke, S. (1975), *J. biol. Chem.*, **250**, 5459–5469.
14. Cochrane, D.E. and Douglas, W.W. (1974), *Proc. natn. Acad. Sci., U.S.A.*, **71**, 408–412.
15. Conrad, D.H., Bazin, H., Sehon, A.H. and Froese, A. (1975), *J. Immunol.*, **114**, 1688–1691.
16. Conrad, D.H. and Froese, A. (1976), *J. Immunol.*, **116**, 319–326.
17. Conrad, D.H., Berczi, I. and Froese, A. (1976), *Immunochem.*, **13**, 329–332.
18. Coombs, J.W., Lagunoff, D. and Benditt, E.P. (1965), *J. Cell Biol.*, **25**, 577–592.
19. Cooper, S.M. and Sambray, Y. (1976), *J. Immunol.*, **117**, 511–517.
20. Coutinho, A., Gronowicz, E. and Möller, G. (1975), In: *Immune Recognition*, (Rosenthal, A.S., ed.), Academic Press, New York pp. 63–83.
21. Davies, G.E. and Stark, G.R. (1970), *Proc. natn. Acad. Sci., U.S.A.*, **66**, 651–656.
22. Day, R.P. (1972), *Clin. Allergy*, **3**, 205–212.
23. Day, L.A., Sturtevant, J.M. and Singer, S.J. (1963), *Ann. N.Y. Acad. Sci.*, **103**, 611–625.
24. DeLisi, C. and Metzger, H. (1976), *Immunol. Comm.*, **5**, 417–436.
25. DeWeck, A.L., Schneider, C.H., Spengler, H., Toffler, O. and Lazary, S. (1973), In: *Mechanisms in Allergy*, (Goodfriend, L., Sehon, A.M. and Orange, R.P. eds.), Marcel Dekker, Inc., New York pp. 323–328.

26. Dickler, H.B. (1977), *Adv. Immunol.*, (In press).
27. Dorrington, K.J. and Bennich, H. (1973), *J. biol. Chem.*, **248**, 8378–8384.
28. Dorrington, K.J. (1976), *Immunol. Comm.*, **5**, 263–280.
29. Douglas, W.W. (1974), In: *Calcium and Cell Regulation*, (Smellie, R.M.S., ed.), The Biochemistry Society, London pp. 1–28.
30. Eccleston, E., Leonard, B.J., Lowe, J. and Welford, H. (1973), *Nature*, **244**, 73–76.
31. Edelstein, S.J. and Schachman, H.K. (1967), *J. biol. Chem.*, **242**, 306–311.
32. Edidin, M., Zagyansky, Y. and Lardner, T.J. (1976), *Science*, **197**, 466–468.
33. Froese, A. (1976), *Immunol. Comm.*, **5**, 437–453.
34. Fujita, Y., Rubinstein, E., Greco, D.B., Reisman, R.E. and Arbesman, C.E. (1975), *Int. Arch. Allergy appl. Immun.*, **48**, 577–583.
35. Gonzales-Molina, A. and Spiegelberg, H.L. (1976), *Fed. Proc.*, **35**, 809.
36. Grefrath, S.P. and Reynolds, J.A. (1974), *Proc. natn. Acad. Sci., U.S.A.*, **71**, 3913–3916.
37. Guyer, R.L., Koshland, M.E. and Knopf, P.M. (1976), *J. Immunol.*, **117**, 587–593.
38. Halper, J. and Metzger, H. (1976), *Immunochem.*, (In press).
39. Hamburger, R.N. (1975), *Science*, **189**, 389–390.
40. Hopf, U., Meyer, K.H., Buschenfelde, Z. and Dierich, M.P. (1976), *J. Immunol.*, **117**, 639–645.
41. Isersky, C., Kulczycki, Jr., A. and Metzger, H. (1974), *J. Immunol.*, **112**, 1909–1919.
42. Isersky, C., Metzger, H. and Buell, D. (1975), *J. exp. Med.*, **141**, 1147–1162.
43. Isersky, C., Mendoza, G. and Metzger, H. (1977), *J. Immunol.*, (In press).
44. Ishizaka, K. and Ishizaka, T. (1969), *J. Immunol.*, **103**, 588–595.
45. Ishizaka, K., Ishizaka, T. and Lee, E.H. (1970), *Immunochem.*, **7**, 687–702.
46. Ishizaka, K., Tomioka, H. and Ishizaka, T. (1970), *J. Immunol.*, **105**, 1459–1467.
47. Ishizaka, T., Soto, C.S. and Ishizaka, K. (1973), *J. Immunol.*, **111**, 500–511.
48. Ishizaka, T. and Ishizaka, K. (1974), *J. Immunol.*, **112**, 1078–1084.
49. Ishizaka, T. and Ishizaka, K. (1975), *Ann. N.Y. Acad. Sci.*, **254**, 462–475.
50. Ishizaka, T., Konig, W., Kurata, M., Mauser, L. and Ishizaka, K. (1975), *J. Immunol.*, **115**, 1078–1083.
51. Johansson, S.G.O., Mellbin, T. and Vahlquist, B. (1968), *Lancet*, **1**, 1118–1121.
52. Kaliner, M. and Austen, K.F. (1973), *J. exp. Med.*, **138**, 1077–1094.
53. Kanno, T., Cochrane, D.E. and Douglas, W.W. (1973), *Can. J. Physiol. Pharmacol.*, **51**, 1001–1004.

54. Keller, R., Peitchel, R., Goldman, J.N. and Goldman, M. (1976), *J. Immunol.*, **116**, 772–777.
55. Konig, W. and Ishizaka, K. (1974), *J. Immunol.*, **113**, 1237–1245.
56. Kulczycki, Jr., A. and Metzger, H. (1974), *J. exp. Med.*, **140**, 1676–1695.
57. Kulczycki, Jr., A., Isersky, C. and Metzger, H. (1974), *J. exp. Med.*, **139**, 600–616.
58. Kulczycki, Jr., A., McNearney, T.A. and Parker, C.W. (1976), *J. Immunol.*, **117**, 661–665.
59. Lawson, D., Fewtrell, C., Gomperts, B. and Raff, M.C. (1975), *J. exp. Med.*, **142**, 391–402.
60. Lay, W.H. and Nussenzweig, V. (1968), *J. Immunol.*, **102**, 1172–1178.
61. Lehrer, S.B. (1977), *Immunol.*, (In press).
62. Leslie, R.G. and Cohen, S. (1974), *Immunol.*, **27**, 577–587.
63. Loor, F., Forni, L. and Pernis, B. (1972), *Eur. J. Immunol.*, **2**, 203–212.
64. Magro, A.M. and Alexander, A. (1974a), *J. Immunol.*, **112**, 1757–1761.
65. Magro, A.M. and Alexander, A. (1974b), *J. Immunol.*, **112**, 1762–1765.
66. Matthysens, G.E., Hurwitz, E., Givol, D. and Sela, M. (1975), *Molec. Cell Biochem.*, **7**, 119–126.
67. Mendoza, G. and Metzger, H. (1976), *J. Immunol.*, **117**, 1573–1578.
68. Mendoza, G. and Metzger, H. (1976), *Nature*, **264**, 548–550.
69. Messner, R.P. and Jelinek, J. (1970), *J. clin. Invest.*, **49**, 2165–2171.
70. Metzger, H. (1976), In: *Cell Membrane Receptors for Viruses, Antigens and Antibodies, Polypeptide Hormones and Small Molecules*, (Beers, R.F., Jr. and Basset, E.G., eds.), Raven Press, New York pp. 289–301.
71. Metzger, H., Budman, D. and Lucky, P. (1976), *Immunochem.*, **13**, 417–423.
72. Metzger, H., Elson, E.L., Isersky, C., Mendoza, G., Newman, S.A., Rossi, G. and Schlessinger, J. (1977), in *Proc. IX International Congress of Allergology*, Buenos Aires, Argentine, 1976, Excerpta Medica, Amsterdam, (In press).
73. Meunier, J.C., Olsen, R.W. and Changeux, J.P. (1972), *F.E.B.S. Letters.*, **24**, 63–68.
74. Minard, P. and Levy, D. (1972), *J. Immunol.*, **109**, 887–890.
75. Mossmann, H., Meyer-Delius, M., Votisch, U., Kickhofen, B. and Hammer, D.K. (1974), *J. exp. Med.*, **140**, 1468–1481.
76. Newman, S.A., Rossi, G. and Metzger, H. (1977), *Proc. natn. Acad. Sci., U.S.A.*, (In press).
77. Ogilvie, B.M. and Jones, V.E. (1968), In: *Cellular and Humoral Mechanisms in Anaphylaxis and Allergy*, (Moral, H.Z., ed.), Karger Basel, New York pp. 13–22.
78. Phillips-Quagliata, J.M., Levine, B.B., Quagliata, F. and Uhr, J.W. (1971), *J. exp. Med.*, **133**, 589–601.
79. Plaut, M., Lichtenstein, L.M. and Bloch, K.J. (1973), *J. Immunol.*, **111**, 1022–1030.

80. Raffel, S. (1973), *Mechanisms in Allergy,* Marcel Dekker, Inc., New York.
81. Rask, L., Klareskog, L., Ostberg, L. and Peterson, P.A. (1975), *Nature,* **257**, 231–233.
82. Rapp, H.J. and Borsos, T. (1970), *Molecular Basis of Complement Action,* Appleton, New York.
83. Rosenstreich, D.L. and Blumenthal,, R.P. (1977), *J. Immunol.,* **118**, 129–136.
84. Rossi, G., Newman, S.A. and Metzger, H (1977), *J. biol. Chem.,* (In press).
85. Schlessinger, J., Koppel, D.E., Axelrod, D., Jacobson, K., Webb, W.W. and Elson, E.L. (1976), *Proc. natn. Acad. Sci., U.S.A.,* **73**, 2409–2413.
86. Schlessinger, J., Webb, W.W., Elson, E.L. and Metzger, H. (1976), *Nature,* **264**, 550–552.
87. Segal, D.M. and Hurwitz, E. (1977), *J. Immunol.,* (In press).
88. Siraganian, R.P. (1974), *Analyt. Biochem.,* **57**, 383–394.
89. Siraganian, R.P., Hook, W.A. and Levine, B.B. (1975), *Immunochem.,* **12,** 149–157.
90. Siraganian, R.P., Kulczycki, Jr., A., Mendoza, G. and Metzger, H. (1975), *J. Immunol.,* **115**, 1599–1602.
91. Stanworth, D.R. Humphrey, J.H., Bennich, H. and Johansson, S.G.O. (1968), *Lancet,* **2**, 17–18.
92. Sullivan, A.L., Grimley, P.M. and Metzger, H. (1971), *J. exp. Med.,* **134**, 1403–1416.
93. Sullivan, T.J., Parker, K.L., Stenson, W. and Parker, C.W. (1975), *J. Immunol.,* **114**, 1473–1479.
94. Taylor, R.B., Duffus, W.P.H., Raff, M.C. and de Petris, S. (1971), *Nature New Biol.,* **233**, 225–229.
95. Tiggelaar, R.E., Vaz, N.M. and Ovary, Z. (1971), *J. Immunol.,* **106**, 661–672.
96. Unkeless, J.C. and Eisen, H.N. (1975), *J. exp. Med.,* **142**, 1520–1533.
96a. Uvnäs, B. (1973), In: *Mechanisms in Allergy,* (Goodfirend, L., Sehon, A.M. and Orange, R.P., eds.), Marcel Dekker, Inc., New York.
97. Vardinon, N., Spirer, Z., Fridkin, M. and Schwartz, J. (1976), Presented at Third European Immunology Meeting, Copenhagen. (Abs. P139).
98. Vijayanagar, H.M., Attallah, N.A., David, M.F. and Sehon, A.H. (1974), *Int. Arch Allergy,* **46**, 375–392.
99. Yiu, S.H. and Froese, A. (1976), *J. Immunol.,* **117**, 269–273.
100. Yoshida, T.O. and Andersson, B. (1972), *Scan. J. Immunol,* **1**, 401–408.

3

Endocytosis

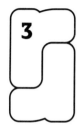

THOMAS P. STOSSEL
Medical Oncology Unit,
Massachusetts General Hospital
and
The Department of Medicine,
Harvard Medical School,
Boston, Mass, 02114, U.S.A.

Receptors and Recognition, Series A, Volume 4
Edited by P. Cuatrecasas and M.F. Greaves
Published in 1977 by Chapman and Hall, 11 New Fetter Lane, London EC4P 4EE

INTRODUCTION

Phagocytosis is the incorporation of relatively large solid objects by cells such that the objects come to reside within intracellular vacuoles [1]. *Pinocytosis* is the imbibition of extracellular fluid in small membrane-bound vesicles [2]. Phagocytosis and pinocytosis are both examples of *endocytosis* in that they involve membrane internalization. Phagocytosis is highly relevant to the alimentation of protozoa and for the activities of the polymorphonuclear leukocytes and mononuclear phagocytes, vertebrate cells that specialize in host defense against microbial infections [3]. However, phagocytosis is a capability of many other cell types. Thyroid cells ingest colloid [4, 5], epidermal cells engulf melanosomes [6] and retinal epithelial cells interiorize cone segments [7], examples of phagocytosis as an activity relevant for tissue functions. Other cells such as fibroblasts [8], tumor cells [9], plasma cells [10], muscle cells [11] and probably many other cells can phagocytize under certain circumstances. Pinocytosis occurs regularly in even more cell types than phagocytosis and is pertinent for diverse cytologic functions ranging from secretion to nutrition.

Phagocytosis and pinocytosis occur when a perturbation of the external cell surface evokes the proper responses in the plasma membrane and subplasmalemmal cortical cytoplasm. Endocytosis is therefore a prime example of cell surface-to-cytoplasm communication, and *recognition* is the first step in the process. The recognition signal presumably elicits a transduction mechanism (or second messenger) which evokes a number of *effector* steps. Some of these effector events are visible to the microscopist. In the case of phagocytosis, contact of a recognizable particle* with the plasma membrane causes the membrane to increase its stickiness, such that it binds firmly to the particle. A region of cytoplasm adjacent to the particle enlarges, and pseudopodia emerge. The pseudopodia appear clear (or hyaline) in the phase contrast microscope, and exclude phase-dense organelles such as the nucleus and granules, and move around the particle, fusing at its distal pole.

* Objects interiorized by cells by means of phagocytosis will be referred to as 'particles.'

Initial contact of the cell and particle may occasionally occur via spidery filipodia which coalesce to create pseudopodial veils [12]. In summary, the effector mechanisms of phagocytosis that are visible include adhesion, pseudopod assembly, pseudopod movement and pseudopod fusion, which recognition elicits. Pinocytosis appears to be mere membrane invagination without pseudopod evagination, and fusion at the neck of the vesicle closes off the pinosome. It is not clear whether the invagination of membrane in phagocytosis is analogous to that which occurs in pinocytosis, or is simply the passive result of the formation of lateral pseudopodia.

One approach to an analytical discussion of endocytosis is to begin as the cell does, with recognition, and to proceed with the transduction and effector mechanisms. However, unlike many other recognition–effector systems, relatively little is known about the initial events of the effector system. It is a relatively simple matter to deal sequentially with the recognition, transduction and effector mechanisms of many endocrine cells, since the effector enzymes have been well characterized, but only the vaguest notion of what controls the responses of endocytosis outlined above is beginning to emerge. The mechanism of pinocytosis is even less clear than that of phagocytosis. Therefore, this review will discuss the embryonic information about effector mechanisms first, setting up some candidates for activation of that mechanism. It will then work backwards through transduction and recognition. The discussion will limit itself to the mechanism of endocytosis, leaving aside the important subject of the fate of the interiorized membrane.

3.1 METHODOLOGIC CONSIDERATIONS

Although there is the distinct advantage of being able to observe, by means of the light microscope, phagocytes ingesting, the interpretation of what is observed is not so simple. The greatest difficulty concerns the differentiation between non-specific attachment of particles to cell surfaces from actual internalization of the particles by the cells, especially when the light microscopist ceases to record simple observations with vital preparations and attempts to quantitate numbers of particles 'ingested' on fixed smears, which is impossible. However, the quantitation of ingestion is essential to the study of phagocytosis. In 1921, Fenn pointed out the importance of measuring the 'velocity' of particle uptake and outlined some of the factors liable to influence rates of ingestion, such as particle concentration and the method for mixing particles and cells [13]. He also differentiated between phagocytic capacity and phagocytic *rate*, demonstrated that the invariable clumping of phagocytes which occurs after prolonged ingestion affects the cell particle interaction, and attempted to design experiments to limit these variables.

A great number of systems have subsequently been devised for assaying ingestion. The morphologic estimation of ingestion, despite its limitations, can be used as a check of or in combination with other more quantitative techniques. Then it is helpful if smears of cells can be treated with chemicals that lyse extracellular particles, leaving intracellular ones as lucent vacuoles [14, 15]. Electron microscopy is often useful in documenting actual particle internalization, but a thin section showing a particle encased within a membrane-bound vacuole is not sufficient evidence for true ingestion unless serial sections reveal that all sides of the particle are truly enclosed within vacuolar membrane. Chambers has utilized ruthenium red to simplify the electron microscopic approach to the ingestion of one cell by another. The cation binds to external cell surfaces where it is visible as an electron-dense band on the cell coat. If incubated with mixtures of phagocytizing cells, it appears on the surface of partially engulfed but not fully ingested cells [16].

Theoretically, the best assays for ingestion involve direct measurement in phagocytes of particles which, either by their intrinsic nature or by being labeled with radioactivity or a dye, permit detection by chemical techniques. Systems of this type have been devised with polystyrene or polyvinyltoluene particles [17, 18], radioactively labeled bacteria [19], starch particles [19], immune complexes [20, 21], or erythrocytes [22], and with paraffin oil particles containing oil red O [23]. The major technical hurdle in these and other assays is efficient separation of cells containing particles from un-ingested particles, usually accomplished by washing cells attached to glass or plastic surfaces in monolayers, or by differential centrifugation of cells and particles in suspension. The types of control experiments which indicate that the separation process is acceptable include:

(1) demonstration that the rate of particle ingestion is saturable with respect to particle concentration – that is, that increasing particle loads do not merely pack on the surfaces of and between clumps of cells, but that a maximum rate of true internalization is achieved;

(2) demonstration of *complete* inhibition of particle ingestion at ice-bath temperature or in the presence of sufficiently high concentrations of

metabolic inhibitors;
(3) demonstration of lack of apparent ingestion at zero time – when the incubation is initiated at physiological temperatures. When measured, high zero time or ice-bath temperature 'ingestion' values have been observed for polystyrene particles [24], paraffin oil particles coated with certain proteins or starch [25], and radioactively labeled bacteria [26]. While it is often assumed that such values can be subtracted from those achieved during incubations at physiological temperatures and at later times, it is possible that at higher temperatures more 'attachment' sites are exposed because of cellular spreading. With other kinds of particles, the zero time and low temperature blanks are low, thereby facilitating measurement of overall ingestion rates [25, 27]. When the artifacts are minimized, it is possible, with quantitative techniques, to measure the increase in cell-associated particles (rather than the decrease in concentration in the extracellular medium), permitting quantitation of the *initial rate* of ingestion. The importance of initial rate measurements in enzyme kinetics and transport phenomena and their utility in studies of ingestion is well established.

The measurement of pinocytosis has been accomplished by counting numbers of vacuoles formed and visible through the light microscope, by assaying the rate of accumulation of labeled matter in the extracellular medium by the cell, or by quantifying the disappearance of labeled material from the medium [101, 104, 191].

3.2 THE EFFECTOR MECHANISM OF ENDOCYTOSIS

3.2.1 Endocytosis and the plasma membrane

The fact that the initial rates of phagocytosis [19, 23, 25] and pinocytosis [104, 179] are constant for a finite period of time [19, 23, 25] indicates that availability of interiorizable membrane is not a limiting factor during this initial period. One explanation for continuous membrane availability is resynthesis that is commensurate with removal. Membrane renewal could result from *de novo* macromolecule synthesis, incorporation of preformed building blocks, or even insertion of whole membrane segments. Evidence for new net lipid synthesis as an acute consequence of phagocytosis has not been impressive [28–30], and phagocytosis proceeds efficiently in the presence of inhibitors of protein synthesis [31–33]. The results suggest strongly that *de novo* membrane synthesis is not an integral aspect of the phagocytic effector

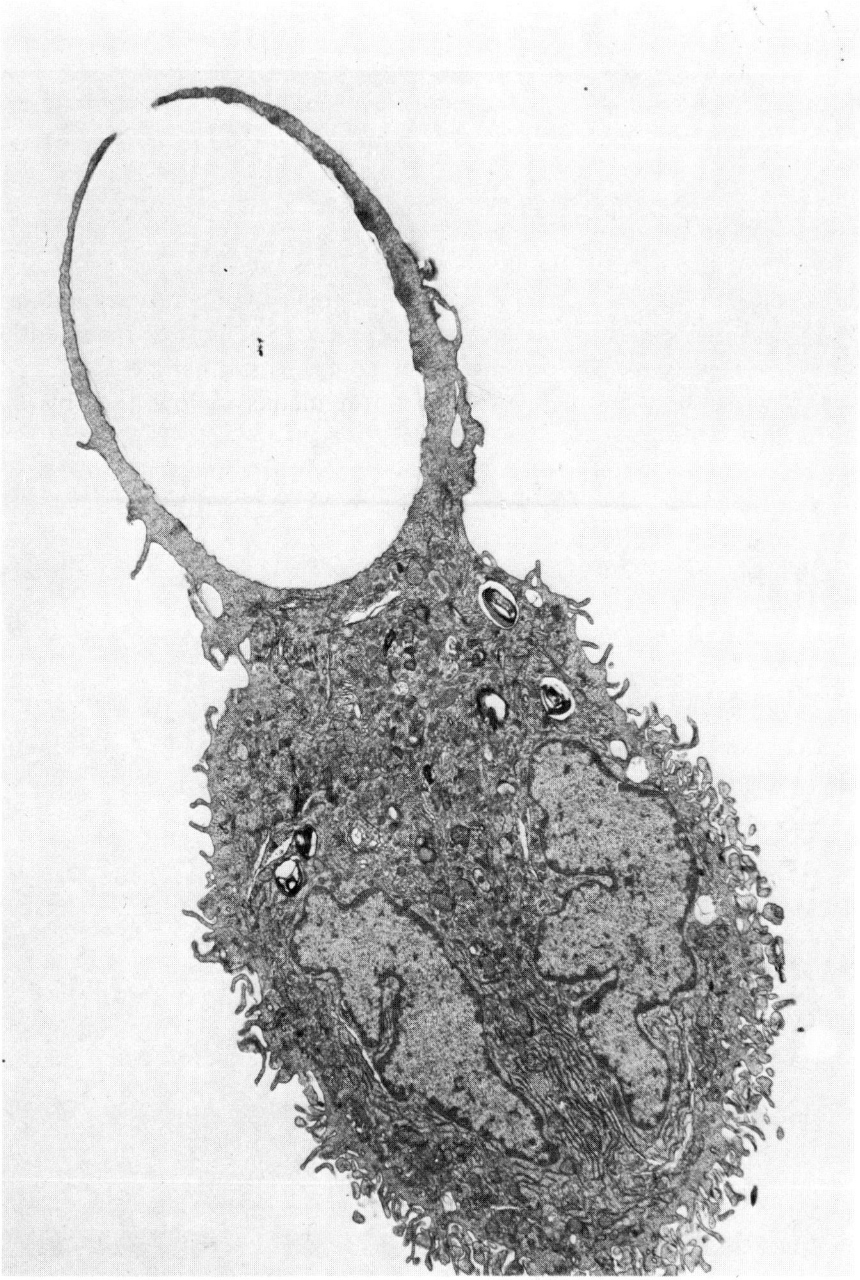

Fig. 3.1 Electron micrograph of rabbit pulmonary macrophage ingesting an *Escherichia coli* lipopolysaccharide-coated oil droplet reacted with the opsonic fragment of C3. Note the membrane redundancy expressed in multiple small pseudopodia that is attentuated in the region of the particle being internalized. Note also the exclusion of organelles from the pseudopodia. (Magnification: 690 x).

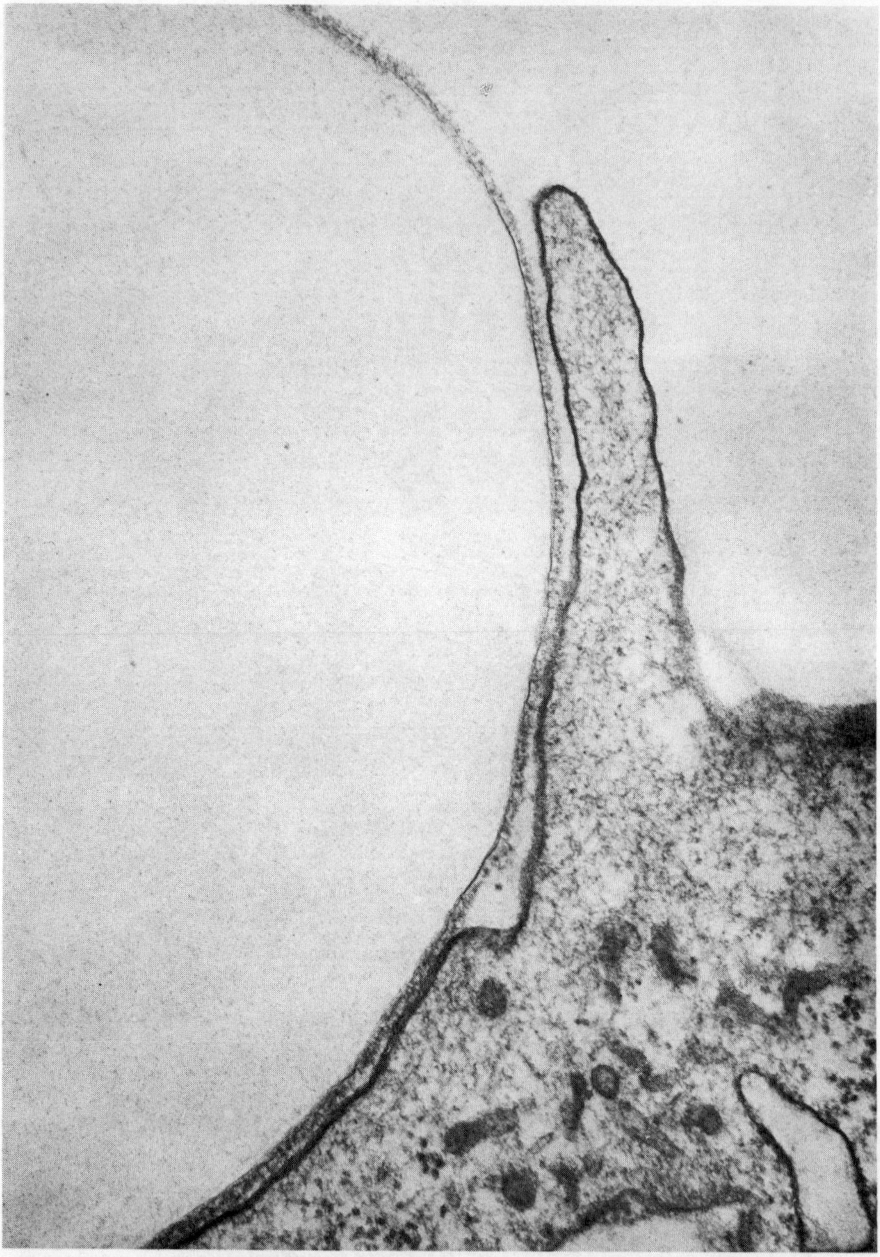

Fig. 3.2 Electron micrograph of the region of a rabbit pulmonary macrophage attached to an *Escherichia coli* lipopolysaccharide-coated oild droplet reacted with the opsonic fragment of C3. Note the randomly oriented thin filaments in the submembrane region and filling the pseudopod which is beginning to surround the particle. (Magnification: 56 600 x).

mechanism. On the other hand, inhibition of protein synthesis impairs pinocytosis [32]. Few complementary studies of macromolecular turnover during pinocytosis are available on which to base an appraisal of the meaning of this inhibition, but the suggestion that membrane recycling occurs during pinocytosis is plausible [192].

Although acute replacement of membrane during phagocytosis from preformed constituents is difficult to exclude, limited analyses of the specific radioactivity of phagocytes or phagocytic membranes, labeled with lipid or protein precursors, do not suggest losses in specific activity which might suggest replacement of membrane by a pool of precursors [33–36].

The most reasonable mechanism to account for the initial linearity of endocytic rates is redundancy of the plasma membrane which allows for extensive losses without functional depletion. Scanning and transmission electron micrographs clearly portray such redundancy (Fig. 3.1) [37, 38].

3.2.2 Evolution of evidence for the role of contractile proteins in phagocytosis

Early microscopists perceived that the assembly and movement of organelle-excluding hyaline ectoplasm was relevant to phagocytosis. From their visual impressions and simple experiments, they concluded that changes in the viscosity of an ectoplasm gel could regulate the form and consistency of cell surfaces, and the term 'contractile' appears in this early work [39]. These astute notions had to remain vague until the advent of techniques for defining the ultrastructure and biochemical composition of the cell periphery.

Although it was apparent in many early thin section electron micrographs that a tangle of 4–6 nm microfilaments, and little else, filled the pseudopods embracing particles undergoing ingestion (Fig. 3.2), this fact became a focus of attention only later [40–42]. As amplified below, it was evident by the late 1960's that these microfilaments were probably actin polymers and that some kind of contractile process had something to do with phagocytosis [43]. This idea received a stimulus from the discovery that the fungal metabolite, cytochalasin B, which was believed in some loose sense to inhibit 'microfilament–contractile' processes [44], inhibited phagocytosis by mouse peritoneal macrophages and polymorphonuclear leukocytes [43, 45–50].

3.2.3 Presence and interactions of contractile proteins in phagocytes

The definitive isolation of the rudiments of a contractile system, actin [51] and myosin [52] in non-muscle cells, opened the way for a biochemical definition of the role of contractile proteins in cellular events. The first wave of investigations concentrated on identifying and purifying these molecules in and from cells, and succeeded in some instances in isolating both actins and myosins with strikingly similar subunit composition and functional properties such as muscle actin and myosin [53]. Although many of these cell types are highly motile cells capable of undergoing phagocytosis at times, the proteins are clearly ubiquitous and present in cells such as platelets and sperm, not known for their phagocytic capability. Therefore, the mere presence of actin and myosin in a cell does not prove that they have anything to do with phagocytosis. Reassuringly, however, all phagocytic cells studied to date, appear to possess ample amounts of actin (up to 10% of the total cell protein) and associated contractile proteins.

Some understanding of how these non-muscle or 'cytoplasmic' muscle proteins work is essential in order to learn how they might relate to phagocytosis. The structural resemblance between muscle and cytoplasmic actins and myosins makes it reasonable to assume that a similar homology might obtain between the mechanism of actin–myosin interactions involved in skeletal muscle contraction and in non-muscle cell activity such as phagocytosis. Briefly, this mechanism, as currently understood, involves the sliding of interdigitating parallel actin and myosin filaments. The cyclic establishment and breakdown between the filaments of cross bridges, which are the head groups on the heavy chains of the myosin molecules protruding from the myosin filaments, effect this sliding. The cross bridge movement is the expression of enzymatic activity in the myosin head, which is a magnesium adenosine triphosphatase (Mg^{2+} ATPase) that is active under physiological conditions of pH and ionic strength, and which is activated by actin polymers. This being the case, a control mechanism must prevent continuous Mg^{2+} ATPase activity and contraction. In skeletal muscle, the control proteins, tropomyosin and the troponins, fulfill this requirement, inhibiting the enzymatic and mechanical actin–myosin interactions. In certain molluscan muscles and in vertebral smooth muscle, the control proteins are part of the myosin molecule, and, in still other muscles, both types of regulation obtain. In all these cases, free calcium ions bind to the regulatory proteins and release the inhibitory effect of these proteins on the actomyosin interaction. Hence, calcium is the ultimate regulator of the contractile activity [54].

Therefore, magnesium ATPase activity, mechanical movement and some control mechanism possibly linked to calcium ions should characterize the actin–myosin interaction of phagocytes. The Mg^{2+} ATPase activity reported for many cytoplasmic actomyosins has been considerably lower than that of skeletal muscle [53], a problem which complicates the study of regulation. For example, an increase in very low Mg^{2+} ATPase activities of crude actomyosin preparations by addition of calcium ions could reflect the recruitment of irrelevant contaminating ATPases [55]. Initially, this meager activity might have been ascribable to the presumably small forces required to be generated, since some proportionality between rates of muscle contraction and of myosin enzymatic activities is demonstrable [56]. Proteolysis during isolation is another possible explanation for inactive actomyosins. However, improved purification schemes for myosin and the discovery of novel co-factors have demonstrated that actomyosin Mg^{2+} ATPase activity of phagocytic cells can be in the range associated with skeletal muscle actomyosin [57–60]. These higher activities set the stage for evaluation of what regulates contractile activity in phagocytic cells, since the ATP hydrolysis rates permit unambiguous interpretation of assay results. The higher activities also force the student of cytoplasmic actomyosin not to accept low Mg^{2+} ATPase activities but to search for factors or conditions that increase it. The actomyosin co-factors discovered in *Acanthamoeba* and rabbit pulmonary macrophages appear to be proteins of about 70 000–90 000 molecular weight and activate the Mg^{2+} ATPase activities of actomyosin in a concentration-dependent manner [57, 60]. Little more is known. The rabbit pulmonary macrophage co-factor also does not appear to act by phosphorylation of myosin, a mechanism postulated for the regulation of contractile activity in platelets and smooth muscle [61, 62]. The calcium concentration does not influence the Mg^{2+} ATPase activity of co-factor-activated *Acanthamoeba* or rabbit pulmonary macrophage actomyosins, either in mixtures of the purified proteins or in relatively crude fractions containing these proteins [60, 63]. On the other hand, calcium is reported to increase the Mg^{2+} ATPase activity of crude *Physarum polycephalum* actomyosin [64], and a preliminary publication advertised the existence of a calcium-sensitizing fraction from *Dictyostelium discoideum* actomyosin [65]. Neither of these species apparently require co-factors for activation of their actomyosin Mg^{2+} ATPase activities [58, 59].

In summary, the rudiments of a contractile system, actin and myosin, exist in phagocytes and in some examples interact enzymatically in a manner and magnitude consistent with the generation of mechanical force. The mechanism of control of that interaction may be somewhat unusual with respect to skeletal muscle.* Whether calcium ions play a role is unclear.

3.2.4 Interactions of contractile proteins and phagocytosis

There are distinct differences between the 'musculature' of phagocytes and of muscle cells. The wispy, disorganized microfilaments of pseudopods (Fig. 3.2) bear little resemblance to the interdigitating parallel thick and thin filaments of skeletal muscle cell. Artefacts of fixation and processing for electron microscopy could in part account for this [66], but not for the argument that it is illogical for a phagocyte to have a highly organized static contractile apparatus, when it must assemble and disassemble pseudopods at random sites on its surface. Furthermore, pseudopodia are hyaline in phase contrast and isotropic viewed under crossed polaroids, which is not consistent with organized fibrils. Impressions from electron micrographs indicate that actin filaments contact the inner surface of the plasmalemma of amoebas [67], macrophages [43, 68], and polymorphonuclear leukocytes [69]. Isolated peripheral blebs of rabbit pulmonery macrophages, hyaline in the phase microscope, contract when warmed in the presence of magnesium ions [70]. Membrane 'ghosts' derived from amoebas behave in a similar way [71]. The findings indicate that the contraction of 'disorderly' actin in the isolated state may occur in the cell periphery, and that this contraction can cause membrane movement.

Biochemistry

The actin to myosin concentration ratio of phagocytes is much higher than that of muscle cells [53], implying that a different sort of interaction and control mechanism might obtain to effect endocytosis. One of these differing mechanisms is the reversible change in the state of actin. Cytoplasmic extracts prepared in the cold from various amoebas,

* The analysis of cytoplasmic actomyosin has used skeletal muscle proteins as a model because of greater understanding of its basis. In fact, cytoplasmic myosins resemble smooth muscle myosin more than skeletal muscle myosin in subunit composition. Ironically, more is known about some aspects of cytoplasmic contractile proteins than of smooth muscle contractile proteins.

leukocytes and macrophages become viscous when warmed in the presence of ATP, and actin filaments form concomitantly [66, 68, 72, 74]. Temperature influences the monomer/filament equilibrium of muscle actin, low temperature favoring the monomeric or G-state [75]. However, the rapid temperature-dependent changes in actin's state observed in cytoplasmic extracts, only occur when the bound adenine nucleotide of muscle actin is ADP rather than ATP [76]. Therefore, either something is unusual about these cytoplasmic actins, or else some other factor or factors impose this odd behavior. Another odd feature is the 'viscosity' of the warmed cytoplasmic extracts. Rather than the thixotropic sol characteristic of F-actin solutions, the warmed cytoplasms have a rubbery rigidity that weakly resists deformation, that is, they gel. After mechanical agitation, the gels break into chunks and then acquire a viscoelasticity more usual for F-actin [66, 68, 74]. Therefore, three states of actin require consideration: a sol (monomeric, G-actin), a thixotropic liquid (F-actin) and a gel (also F-actin) (Fig. 3.3).

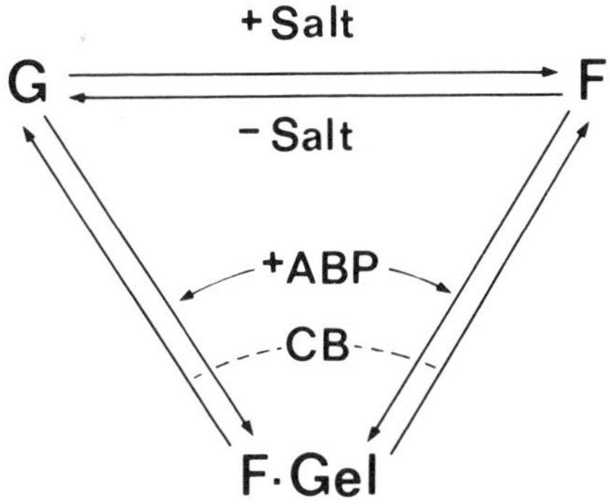

Fig. 3.3 Scheme for possible states of actin in phagocytes: (G), monomeric actin, (F), polymers of actin, (F-Gel), actin polymers cross-linked or rigidified to form a gel. The gel state of F-actin in human polymorphonuclear leukocytes and rabbit pulmonary macrophages depends on an actin-binding protein (ABP). Cytochalasin B (CB) inhibits the gelation of F-actin by actin-binding protein in the presence of salt and inhibits the assembly and gelation of actin in the absence of salt.

Since the conditions under which actin in cytoplasmic extracts is truly monomeric are somewhat unphysiological (absence of monovalent salts or magnesium chloride, and low temperature) the question as to whether much actin is ever truly monomeric in intact phagocytes, is moot. However, lowering of the free calcium concentration of various amoeba

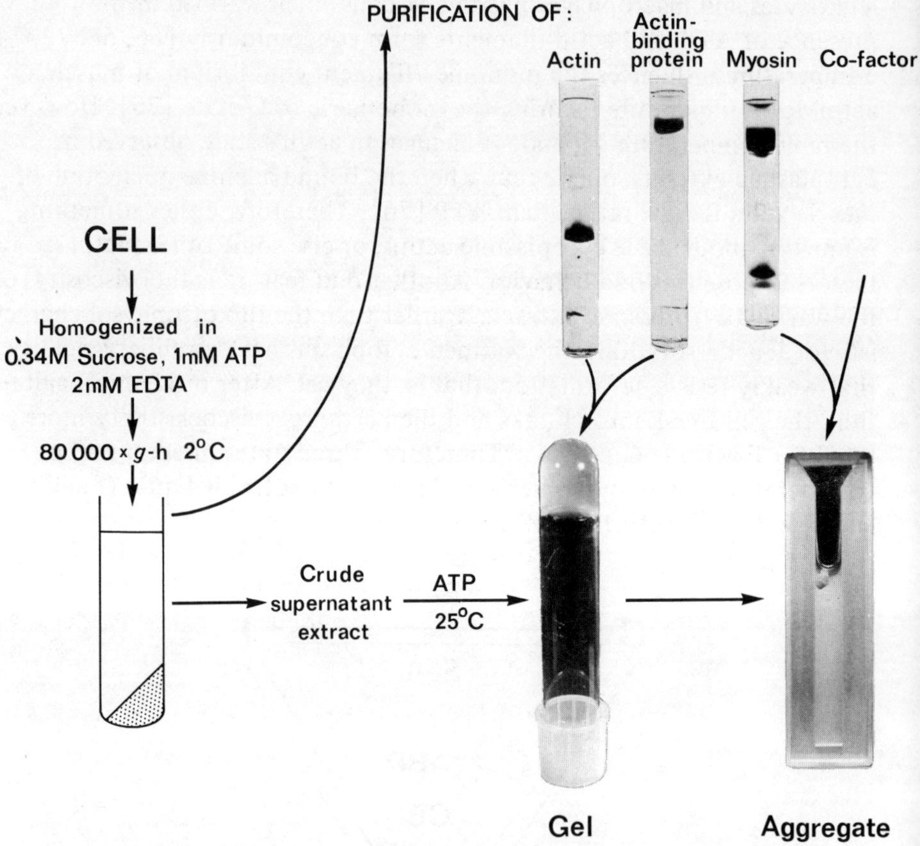

Fig. 3.4 Proteins involved in the gelation and contraction of cytoplasmic extracts of rabbit pulmonary macrophages [169]. The proteins are shown after electrophoresis on 5% polyacrylamide gels with sodium dodecyl sulfate and stained with brilliant coomassie blue.

extracts, maintained at what is otherwise presumed to be normal intracellular ionic composition, caused actin filaments to disappear and birefringence to diminish [71]. If monomer filament interchange were to occur in the cytoplasm of phagocytes, then factors regulating the equilibrium in both directions might exist. Apart from agents already known to influence the aggregation state of actin, such as salts, purine nucleotides and pH [75], new candidates in the form of 'actin-binding proteins' have recently emerged. A high molecular weight (circa 280 000 daltons) protein that has been purified from rabbit pulmonary macrophages [77] and human polymorphonuclear leukocytes [74] is required for the temperature-dependent reversible polymerization of macrophage, leukocyte, and rabbit muscle actins in sucrose solutions containing ATP and low salt concentrations [68, 74] (Figs. 3.3, 3.4). The mechanism by which these actin-binding proteins promote the assembly of actin under these conditions is not known, but other proteins which bind actins, myosin and α-actinin, are known to increase the rate of actin polymerization under defined conditions [78, 79]. In phagocytic amoebas a 280 000-dalton protein is among others present in the extracts in which actin undergoes temperature-dependent reversible assembly [66, 71].

The purified actin-binding proteins of rabbit pulmonary macrophages and human polymorphonuclear leukocytes also promote the rapid temperature-dependent reversible gelation of purified actins [68, 74]. Rabbit muscle actin, rabbit pulmonary macrophage actin and human polymorphonuclear leukocyte actin do not gel at concentrations up to 4 mg ml^{-1} without addition of the actin-binding protein [68, 74], but highly purified *Acanthamoeba* actin was reported to gel slowly on its own [66]. Other factors influencing actin gel formation in mixtures of purified actin and actin-binding protein, or else in crude cytoplasmic extracts, are salts and ATP. Gel formation is maximal under conditions of pH, ATP, magnesium chloride, and potassium chloride concentrations likely to obtain in the cell [71]. Low (10^{-6} to 10^{-8} molar) concentrations of calcium chloride may increase the rate of actin gel formation in cytoplasmic extracts [66, 68, 71, 74], but gels do form at very low calcium concentrations (that is, in the presence of calcium chelators [66, 68, 71, 74]. Therefore, actin-binding proteins, monovalent salts, ATP and possibly calcium are potential physiologic promoters of actin gelation, but no agents which could reverse the gel state in the cell have yet been discovered.

Focal gelation is one way to regulate localized movement. A given quantity of myosin, acting on actin polymers stabilized by cross-links or possibly rigidified by other factors, can move more actin than myosin working on individual filaments. This idea is the fulcrum for the major hypothesis of this discussion.

3.2.5 Focal actin assembly and/or gelation initiates the effector mechanisms of phagocytosis

A corollary to this hypothesis is that the actin state change is a regulator of contractile activity. Experiments with purified proteins indicate that myosin contracts cytoplasmic actin gels [68, 74], and, in the case of rabbit pulmonary macrophage proteins, actomyosin co-factor accelerates

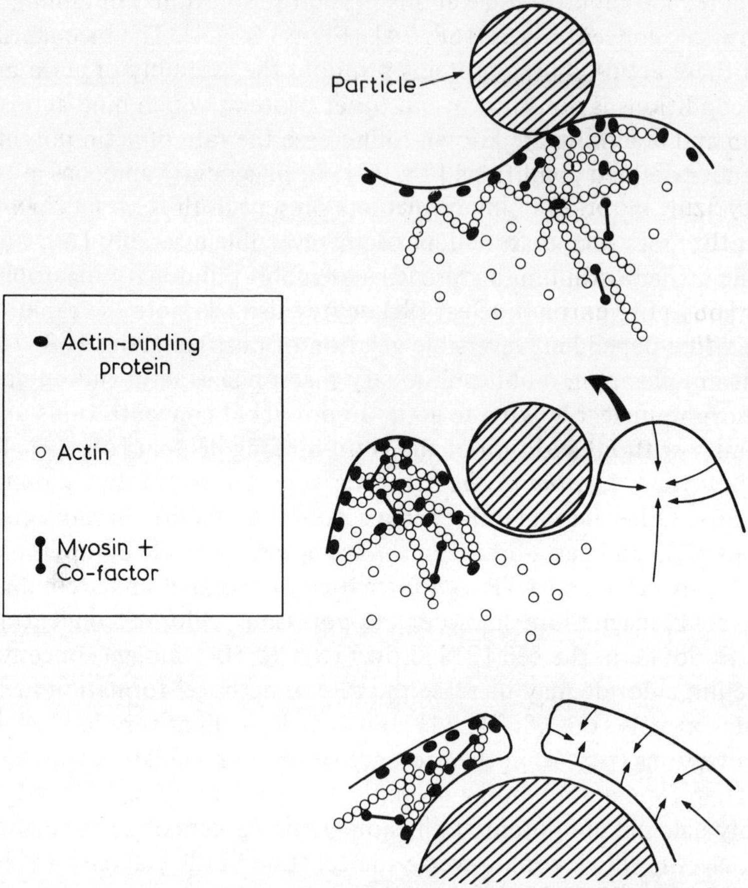

Fig. 3.5 Hypothetical mechanism involving contractile proteins for the engulfment of particles by phagocytes [193]. Contact with an ingestible particle elicits a change in the intracellular actin-binding protein. Alteration of the membrane-association of this protein [194] promotes the assembly and gelation of cytoplasmic actin. Contraction of the actin gel by myosin brings new membrane sites in contact with the particle to propagate the reaction and compresses the membrane to form the speudopod. Sequential activation effects the circumnavigation of the particle by the membrane.

the contraction rate (Fig. 3.4) [68]. Crude gelled cytoplasm of amoebas also contract, and indirect evidence implies that myosins are the molecules responsible for the force generation [66]. According to the model proposed, the availability, cross-linking, and rigidity of actin filaments determine what the available myosin molecules have to work on and thereby, the extent or efficiency of contraction. Factors influencing the Mg^{2+} ATPase activity of the myosin (or the state of activation of co-factors) could impose further regulatory effects on that interaction. The second corollary to the hypothesis is that the assembly and movement of pseudopodia which characterize phagocytosis result from the focal gelation and membrane association of actin filaments. Extracts of phagocytizing rabbit pulmonary macrophages contain more actin-binding protein than extracts of resting, or non-phagocytizing, macrophages, but the same concentration of actin and myosin as extracts of resting macrophages [68]. The extracts of the phagocytizing cells reflect this higher actin-binding protein concentration in having faster rates of gelation. Since the concentrations of actin-binding protein of whole cells do not change as a result of phagocytosis, actin-binding protein either alters its subcellular distribution or becomes more extractable from membranes during phagocytosis. These observations suggest that a state or location change of actin-binding protein may be a primary event in the phagocytosis-induced activation of the contractile mechanism (Fig. 3.5). If true, then actin-binding protein is an internal transducer of the phagocytic contractile mechanism and can be examined as a target for more proximal activating factors. The idea that particle contact also activates myosin Mg^{2+} ATPase needs to be explored as well.

3.2.6 Concomitant impairment of phagocytosis and contractile protein functions

Cytochalasin B

As mentioned previously, there has been a conviction that the fungal product, cytochalasin B, inhibits microfilament function [44]. The concentrations reported at which the drug inhibits phagocytosis have been at a lower limit of 10^{-6} M, (with the maximal effect being observed at 10^{-5} M). At this concentration, inhibition of phagocytosis was total in most studies. Where less than complete inhibition occurred, the question must be asked whether observed 'ingestion' was not an artefact of the assay utilized. 10^{-5} M cytochalasin B has relatively little effect on a number of functions of muscle and cytoplasmic contractile proteins, including the rate of actin assembly in the presence of salt, the rate of binding of myosin to F-actin, and the magnesium ATPase activity of actomyosin [80–83]. However, this concentration of the drug does inhibit markedly the rate of hexose transport by a variety of phagocytic cells [45, 85]. The interference with hexose uptake begins at a con-

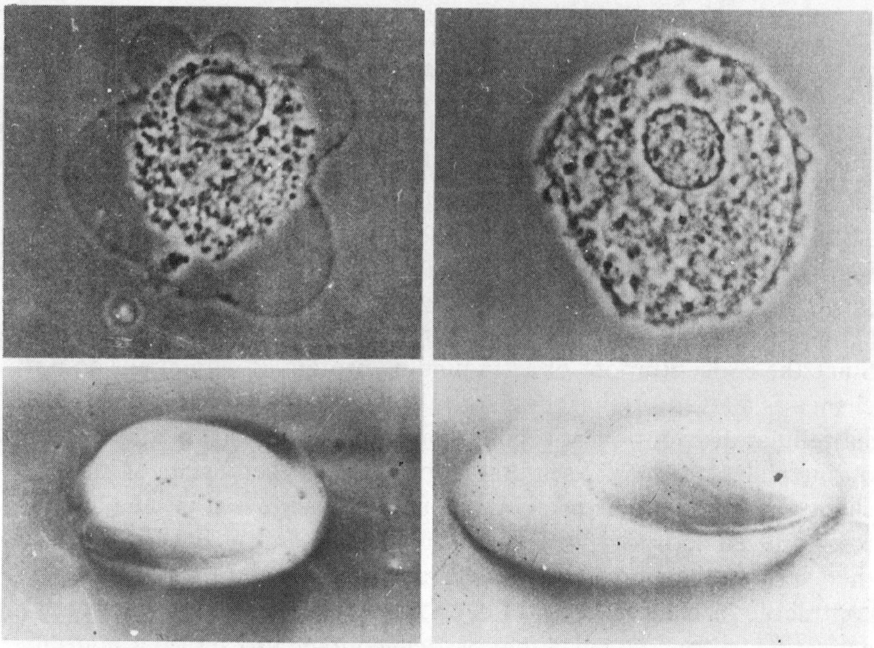

Fig. 3.6 Effect of 10^{-5} M cytochalasin B on the morphology of the hyaline ectoplasm of a rabbit pulmonary macrophage (top) and on the gel state of macrophage cytoplasm (bottom). Macrophage cytoplasmic extract was warmed from 0 to 25°C in a small glass cylinder held upside down on a piece of plastic film. After the extract gelled, the cylinder was removed (left). Addition of cytochalasin B (right) dissolved the gel [86].

centration of 2×10^{-5} M, and is maximal at 10^{-6} M, where inhibition of phagocytosis is not detectable [49]. Moreover, phagocytosis by blood cells occurs with equal efficiency in the presence or absence of added glucose [25, 84]. Therefore, the likelihood that cytochalasin B prevents phagocytosis by acting primarily on membrane glucose transport is remote.

Recently, examination of cytochalasin B's effect on the interaction between actin-binding protein and actin [86], and on actin gelation [86, 87], revealed that the drug dissolves actin-binding protein-actin gels [86] and crude cytoplasmic actin gels [86, 87] (Figs. 3.3, 3.6). These effects of cytochalasin B begin at a concentration of 10^{-7} M, and are maximal by 10^{-6} M, and are reversible [86]. These discoveries indicate that cytochalasin B does not dissolve microfilaments, but rather alters their quaternary structure, and buttresses the relevance of actin-binding protein for actin gelation, and of actin gelation for phagocytosis.

3.2.7 Neutrophil actin dysfunction

The neutrophils of a human infant with recurrent pyogenic infections were markedly deficient in their ability to ingest particles. Actin in extracts of the infant's neutrophils failed to polymerize in the presence of salt concentrations which fully assembled actin in extracts of normal neutrophils [88]. Only half of the actin in extracts of the neutrophils of the patient's parents assembled, suggesting that this disorder of actin assembly was transmitted as an autosomal recessive trait. The description of this remarkable experiment of nature does not prove that actin assembly is required for phagocytosis, but suggests strongly that its failure prevents it.

3.2.8 The mechanism of adhesion in phagocytosis

Light microscopy and scanning and transmission electron micrographs of phagocytizing cells show the plasma membrane apparently tightly adherent to particles being ingested (Figs. 3.1, 3.2). Such adherence provides the frictional force to permit motion of a pseudopod around an object. During ingestion the cell effectively excludes all but minimal amounts of extracellular fluid [18, 89, 90], which supports the notion of a firm barrier created by the bond between particle and cell. During ingestion, phagocytes can become very 'sticky', adhering to one another and to surfaces as well as to particles [91]. Inhibitors of cellular energy metabolism impair the adhesivity of phagocytes, suggesting that the 'sticky' state arises from an active mechanism and not directly from non-specific physical forces existing intrinsically in apposed particles and cells [92, 93] although once established, the adhesive state persists, at least for a short time [94]. The phenomena once again evoke the idea that a process of active rearrangement or metabolism of membrane constituents exposes adhesive sites [95]. The nature of both the process and the sites is mysterious, but invocation of subplasmalemmal contractile activity is reasonable. Cell-to-cell [96], cell-to-substrate [42, 96], as well as cell-to-particle contact sites all have prominent subplasmalemmal microfilament networks, and cytochalasin B can reduce cell-to-substrate adhesion, among its other effects [97]. For movement of the inner membrane face to change the shape, charge and hydrophobicity of the outer face is a reasonable demand. These three physical variables all influence the adhesion of objects to one another [98].

3.3 POSSIBLE ROLE OF CONTRACTILE PROTEINS IN THE EFFECTOR MECHANISM OF PINOCYTOSIS

The morphology of pinocytosis is more difficult to observe than that of phagocytosis because of its smaller scale of dimensions. The suggestion that pinocytosis may involve a cellular contractile mechanism to some extent arises from its association with peripheral membrane movement or ruffling [99], its apparent energy dependence [32], and the observation of thin filaments attached to pinocytic vesicles in the cytoplasm [100]. Pinocytosis differs from phagocytosis in that large hyaline filament-rich pseudopods do not form, and that cytochalasin B does not invariably inhibit it [47, 101, 102], suggesting that cytoplasmic gelation is not always essential for pinocytosis. The metabolic requirements of pinocytosis and phagocytosis also differ, in that poisons of glycolysis which inhibit phagocytosis do not impair pinocytosis unless respiratory poisons are also present [32, 102, 104]. Given the paucity of solid experimental data, rather egregrious speculation is required to explain how contractile events might effect pinocytosis, if at all. One such explanation would have focal contraction of actin filaments pucker the membrane and simultaneously remove peripheral proteins from regions of that membrane. The ensuing loss of membrane-stabilizing influences would then allow the lipid bilayer to acquire a thermodynamically favorable state, a vesicle. Of possible relevance are observations on the behavior of erythrocyte membranes which vesiculate when depleted either generally [105] or locally [106] of associated proteins. Particularly interesting in this regard is the phenomenon of erythrocyte membrane endocytosis which morphologically resembles pinocytosis [107]. The process of endocytosis in erythrocyte ghosts requires divalent cations and ATP [107, 108], thereby resembling myosin-associated contraction. However, actin, but not a true 'myosin', has shown itself to investigators of erythrocyte proteins. If erythrocyte endocytosis is truely 'contractile', then either an actin state change, per se, or else a novel ATP-dependent mechanism of actin movement has to be involved.

3.3.1 The mechanism of membrane fusion in endocytosis

Membrane fusion completes the formation of both phagocytic and pinocytic vesicles. If membrane fusion in this setting is the result of an inherent property of the membranes involved, then removal of a structural barrier between apposed membrane surfaces would permit fusion. On the other hand, membrane fusion could alternatively or additionally require that active changes occur within the membranes at the fusion sites, either involving rearrangement or metabolism of membrane molecules. The only experimental information available concerning these points is quite indirect and only inferentially relevant. There is an extensive phenomenology concerning the extrusion of neutrophil and macrophage lysosomal enzymes into the extracellular medium, an event that involves membrane fusion. Cytochalasin B enhances the secretion of lysosomal enzymes by phagocytes in the presence of other activating factors, for example, ingestible particles or serum containing activated complement fragments [109–113]. The concentrations of cytochalasin B which enhance enzyme release in these systems are in the range which inhibit ingestion by phagocytes, suggesting weakly that the drug acts on the contractile mechanism rather than on the hexose transport mechanism. Independent corroboration of this idea stems from the observation that the neutrophils of the human with dysfunctional neutrophil actin secreted increased amounts of lysosomal enzymes into extracellular fluid and into the few phagosomes formed by these cells relative to normal neutrophils [88]. The studies indicate that the gel or polymerization state of actin can regulate granule-plasmalemmal fusion. On the other hand, no data is available with which to determine whether the change in contractile proteins has only a passive role, permitting membrane-membrane interactions, or whether contractile proteins also actively move membrane moieties, and that the movement is essential for fusion.

In addition to cytochalasin B, other agents promote lysosome plasmalemma fusion, including divalent cations [114], divalent cation ionophores [115], and a staphylococcal toxin called leucocidin [116]. The lectin concanavalin A can inhibit fusion [117]. These substances very likely interact initially with the plasma membrane, but that the mechanism of fusion is a direct consequence of that interaction is unproven.

In summary, so little critical experimentation deals with membrane fusion in endocytosis that to discriminate between limitless theories to explain its mechanism is now impossible except to infer that contractile proteins have some role in regulating it.

3.3.2 Microtubules and endocytosis

The discourse concerning contractile proteins has been lengthy and detailed in proportion to their emerging importance for understanding the mechanism of endocytosis. Considerably less firm information is available about other cytoskeletal structures such as microtubules possibly involved in this event. Electron micrographs of thin sections of mouse peritoneal macrophages [42] and human polymorphonuclear leukocytes [162] show microtubules permeating the region of sub-plasmalemmal microfilaments adjacent to polystyrene particles undergoing ingestion. The microtubules enter the microfilament material in a direction perpendicular to the axis of the plasma membrane. Other evidence of a role for microtubules in phagocytosis is quite indirect. Some but not other studies report inhibition of phagocytosis by antitubulin drugs, colchicine and vinblastine (reviewed in [118]). The uneven quality of assays used in the various analyses obscures the differences.

3.4 THE TRANSDUCTION MECHANISM OF ENDOCYTOSIS

An important factor, aside from knowing the effector systems in deciding the nature of the 'second messenger' of phagocytosis, is to determine whether recognition leads to a highly localized or diffuse response. A diffuse response would favor soluble mediators, such as ionized calcium, or cyclic AMP, or, alternatively, a propagation reaction like the action potential of peripheral nerve. A localized response would not rule out a soluble messenger or propagation reaction, but would require stringent controls to limit their extent. Three lines of evidence suggest that phagocytosis is a localized or segmental reaction. First, mouse peritoneal macrophages readily ingest latex particles or antibody-coated bacteria. As they ingest these particles, they do not interiorize erythrocytes bound to their membranes by means of a sandwich of Fab fragments: univalent rabbit anti-cell Fab attached to mouse peritoneal macrophages and mouse erythrocytes respectively, being connected by bivalent anti-rabbit IgG-Fab. The membrane response to latex or IgG-coated bacteria therefore does not extend to the non-specifically bound erythrocytes. Addition of intact anti-erythrocyte IgG causes the erythrocytes to be ingested, proving that the non-specific binding technique does not block ingestion [119]. Second, mouse peritoneal macrophages 'half-ingest' erythrocytes coated over half of their surfaces with IgG antibody. Addition of further IgG antibody to coat the remainder of the erythrocytes results in their complete ingestion by the macrophages [120]. Third, in a variation of the preceding experiments, mouse peritoneal macrophages fully ingest mouse lymphocytes reacted in the cold with anti-lymphocyte serum. The macrophages bind but do not ingest lymphocytes reacted with antibody at 37°C, which causes the antibody to 'cap' in one region of the cell. Cytochemical techniques demonstrated that the IgG on the lymphocyte is in the region of binding to the macrophage, and electron microscopy showed that subplasmalemmal microfilaments were prominent in the binding zone [121]. The selective nature of the phagocytic response needs consideration in evaluating the 'diffusible' possible transducers discussed below and detracts from their candidacy.

3.4.1 Divalent cations

The fact that divalent cations in the extracellular fluid may promote phagocytosis has been known for many years [122] and reviewed in [118]. The effect of these ions is not uniform or straightforward and depends on the cell type and particle fed to the cell. In some cases, no ingestion occurs in the absence of added divalent ions, and these ions have a concentration-dependent activating effect. Magnesium has an activity on pinocytosis or phagocytosis equal to or greater than that of calcium, and the effects are sometimes additive [84, 123, 125]. In other examples, ingestion occurs in the absence of added ions, and in a few instances even in the presence of EDTA [123, 124]. Although divalent ions promote erythrocyte ghost endocytosis [107, 108], even less information exists as to what divalent cations might have to do with pinocytosis than with phagocytosis. The phenomena suggest that extracellular divalent cations could be recognition co-factors for certain particles or for other endocytosis inducers, rather than second messengers. What effect, if any, the presence or absence of extracellular divalent cations have on the concentration of membrane-associated or free intracellular cations in phagocytes is unknown.

The gelation and contraction of phagocytes' cytoplasm discussed in a previous section require magnesium ions. Given the high concentrations of magnesium ions in most cells, *regulation* of contractile activity by changes in the levels of this ion is difficult although not impossible to countenance. The role of calcium ions in these processes is still uncertain, but, based on the skeletal muscle model, conceivable. At present, the candidacy of divalent ions as endocytosis transducers is still viable but not yet overly compelling.

3.4.2 Cyclic purine nucleotides

Table 3.1 summarizes the criteria which cyclic nucleotides would have to meet to *mediate* endocytosis. Most work addressed to the role of cyclic

Table 3.1 Some criteria to be fulfilled to prove that cyclic purine nucleotides mediate endocytosis.

1.	Ingestible particles or pinocytosis inducers should increase intracellular cyclic nucleotide levels of phagocytes.
2.	A nucleotide cyclase sensitive to ingestible particles or pinocytosis inducers should exist in cell-free membrane preparations of phagocytes.
3.	Cyclic nucleotide phosphodiesterase inhibitors should enhance rates of phagocytosis or pinocytosis.
4.	Added cyclic nucleotides or their derivatives should enhance rates of phagocytosis or pinocytosis.
5.	Agents which increase the intracellular cyclic nucleotide levels of phagocytes should increase the rates of phagocytosis or pinocytosis.

nucleotides and endocytosis has studied the polymorphonuclear leukocytes of various species. Experiments so far have failed to fulfill unambiguously the criteria of Table 3.1. Ingestible particles did not induce rapid increments or decrements in cyclic nucleotide levels, although some exceptions were reported (reviewed in [118]). Particle-sensitive cyclase enzymes have eluded detection. Reports concerning added cyclic nucleotides, their dibutyryl derivatives, or agents known directly or as phosphodiesterase inhibitors to increase cyclic nucleotides indicate that cyclic AMP has modest inhibitory effects on ingestion [118], whereas cyclic GMP enhances it [126] or has no effect [88, 127]. Limited investigation of the role of cyclic nucleotides in phagocytosis by other cell types is no more straightforward. Agents increasing cyclic AMP levels of rabbit pulmonary macrophages were found to inhibit [84] or enhance [128] ingestion; and ingestion either increased very slightly [129] or had no effect [130] on cyclic AMP levels of rabbit pulmonary macrophages. Cyclic AMP, however, did stimulate pinocytosis in toad bladder cells [131] and ingestion in thyroid cells [4], suggesting that cyclic nucleotides may modulate endocytosis in systems where cyclic nucleotides can be clearly shown to regulate an overall cellular response. How cyclic nucleotides might mediate endocytosis is unclear. Speculation is possible concerning a popular idea that phosphorylation states of polymerizable proteins regulated by cyclic nucleotides might affect monomer: polymer equilibrium, but no firm data supports it yet in the context of phagocytosis. If better evidence emerges for phosphorylation of myosin in regulation of contractile activity of phagocytes, the enthusiasm for finding how cyclic nucleotides might control this phosphorylation would heighten.

3.4.3 Proteolysis

The increased extractibility of actin-binding protein in macrophages during phagocytosis raises the question as to how the protein redistribution occurs. One possibility is the cleavage of the molecule by a protease, leading to its detachment from the plasma membrane. Pre-incubation of the serine protease inhibitor, diisopropylfluorophosphate (DFP) with rabbit polymorphonuclear leukocytes has no effect on the ingestion of complement-coated erythrocytes, provided the drug is washed away from the phagocytes prior to addition of the erythrocytes. If the agent is present during exposure of the cells to the particles, ingestion does not occur. Since DFP binds covalently to serine residues, the inference is that ingestion exposes or activates serines not reactive at rest. A profile of DFP analogues is active in this system in proportion to the phosphorylation potential of the compounds [132]. This information is the extent of data available to implicate an 'activatable esterase' in phagocytosis, and aside from the proteolytic transducer mechanism suggested above, the possible importance of this 'esterase' is uncertain.

3.5 THE RECOGNITION MECHANISM OF PHAGOCYTOSIS

Phagocytosis is clearly a process initiated at the external surface of the plasma membrane, since a particle immobilizes its recognition-inciting aspect. Attempts to learn about the physical and chemical characteristics of particles that either do or do not incite recognition by phagocytes and efforts to characterize the surface moieties of phagocytes that respond to these characteristics are the approaches taken to recognition. The ambiguities associated with problems of methodology concerning the quantification of binding and ingestion of particles by phagocytes are most acute and perverse in dealing with the subject of recognition.

3.5.1 The surface properties of particles

'Anti-phagocytic' is a term often applied to surfaces that do not elicit ingestion, for example, the capsules of certain micro-organisms. This term implies that the surface resists recognition actively rather than simply lacking attributes that promote ingestion. Only rarely does information exist to document such an active effect, one example being the cytotoxic effect of streptoccal M protein [133].

Investigators have reported impressive corollations between particle surface charge, hydrophobicity and ingestibility. Enhanced ingestion was attributed to a decrease in net negative surface charge, determined by particle electrophoresis or binding of charged molecules, and to an increase in hydrophobicity, determined by contact angle measurements or oil/water partitioning [134–136]. However, cells have overall net negative surface charges and hydrophobic lipid membranes which render them prone to bind positively charged and hydrophobic objects non-specifically, and this binding may be misinterpreted to represent ingestion. Although at a critical level of interaction charged and hydrophobic forces are certainly relevant to recognition, steric factors may render *net* charge and hydrophobicity irrelevant. For example, as a monolayer of γ-globulin forms on polystyrene particles, the net charge of these particles becomes more positive (less negative) but ingestion of them by human polymorphonuclear leukocytes diminishes markedly [137]. On the other hand, deposition of antibody on tubercle bacilli also decreases their net negative charge but increases their ingestibility by neutrophils [135]. The discrepancy indicates that the steric nature of protein deposition on a particle, e.g. antibody combining with antigen as opposed to binding to a non-specific surface, may have a greater role than the overall charge that it imposes.

Much effort on the subject of recognition has gone into the study of agents in vertebrate serum that coat micro-organisms and other objects, increasing the palatability of those objects to phagocytes (reviewed in [118]). These substances bear the name 'opsonin' (from the Greek: to prepare for dining) given by Wright and Douglas who first described the recognition-promoting effect of serum [138]. Of the opsonins, the IgG antibody molecules and a fragment of the third component of complement (C3) are by far the best characterized. IgG binds specifically to particles or soluble molecules by its Fab combining site, and then reacts with the phagocyte in the $C_{\gamma}3$ domain of its Fc region [139–143].

The sequential interaction of antibody and the classic complement components C1, C4, C2, and C3 is sufficient to opsonize erythrocytes and various micro-organsims. Sera deficient in antibody, C4, C2, or C1, have heat-labile opsonic activity equivalent to that of normal sera, suggesting that the alternative complement pathway which can operate in the absence of antibody is activating C3 and thereby opsonizing as well [27, 147]. Since all of the proteins of the alternative pathway system have yet to be isolated, this idea cannot be tested by demonstrating opsonization with purified components. However, it is strongly supported by the observations that one purified alternative pathway protein, factor B, restores some opsonic activity to heated serum [147] (factor B is heat-labile) and much activity to cord serum deficient in factor B [27].

Addition of ^{125}I-labeled purified C3 to serum results in the fixation of radioactivity to particles as they become opsonized, indicating that opsonization involves fixation of C3 in some form on the surface of particles [27, 145]. The major opsonic fragment of C3 is a 140 000 molecular weight molecule composed of two 70 000 molecular weight chains linked by disulfide bridges. The opsonic fragment of C3 binds very firmly to particles and is extremely resistant, once bound, to physical or chemical denaturation [148]. However, its particle binding site and opsonic effector site remain unknown. Another fragment of C3, smaller than the major opsonic fragment, may promote ingestion under certain circumstances [149]. Although the understanding of the molecular basis of opsonic expression is incomplete, the ability to work with defined substances that promote recognition, creates the opportunity to pursue this understanding.

3.5.2 Ingestible particles and the cell surface

Experiments to elucidate the nature of the interaction between particles and phagocytes have concerned measurement of overall ingestion and attempts to measure particle/cell binding.

Ingestion

Morphological and kinetic studies of polystyrene bead uptake by *Acanthamoeba* indicate that a minimum volume (probably square surface area) of particle must accumulate at the membrane for a segment of that membrane to interiorize [18, 41]. These findings are consistent with the idea that particles must 'trigger' a minimum area of membrane in order for ingestion to occur. Application of this interpretation to investigations of the ingestion kinetics of albumin-coated oil particles by polymorphonuclear leukocytes and macrophages suggests the IgG antibody directed against the albumin on the particle surface reduces the area required for triggering, whereas coating of the albumin with opsonically active C3 does not. Both opsonins increase the overall rate at which the 'triggered' membrane interiorizes [124]. Therefore, IgG may enhance the efficiency of the triggering mechanism as well as its overall rate of activation and would explain why antibody may occasionally be more effective in promoting ingestion than C3, which often only promotes partial ingestion. This incomplete ingestion morphologically

ACTIVATION OF ENDOCYTOSIS IN PHAGOCYTES

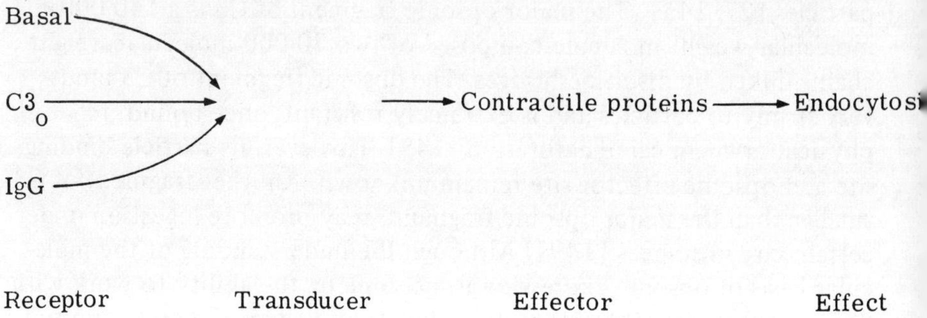

Fig. 3.7 Scheme for the involvement of receptors, transducer, and effector mechanisms in endocytosis.

simulates 'binding' or 'attachment' [150–152].

The most popular interpretation for the selectivity of responses of phagocytes to certain particles is that the cells possess 'receptors' specific for homologous configuration on the particle surfaces, namely receptors for IgG (the Fc receptor), the opsonic fragment of C3 and 'non-specific' or non-immunologic surfaces (Fig. 3.7). Consistent with the breakdown of recognition into distinct 'receptors' is the effect of certain manipulations of the cell on the overall ingestion rate of various particles. For example, the trypsinization of polymorphonuclear leukocytes and macrophages impairs the ability of these cells to engulf C3-coated particles without appreciably reducing their avidity for IgG-coated particles [151, 153, 154]. In some, but not all, experiments, aggregated or free IgG inhibits the ingestion of IgG-coated particles but not of C3-coated or uncoated particles by polymorphonuclear leukocytes or macrophages [139, 152, 153, 155, 156]. Anti-phagocyte antisera have greater inhibitory effects on the uptake of IgG-coated particles than on the ingestion of C3-coated or miscellaneous objects [158–159]. Finally, macrophages which have bound to antibody complexes on part of their surface, will not ingest antibody-coated particles with their remaining free surface although they do engulf C3-coated particles [159]. These experiments suggest engagement of IgG 'receptors' independently of C3 'receptors.'

Since the opsonic effector molecules described above are inhabitants of vertebrate serum of higher animals, the ability of various cell types to respond to these opsonins is intructive in determining whether 'receptors' for these effectors have evolved in parallel with the opsonins or whether the opsonins developed to accomodate a general recognition mechanism of phagocytes. The answer is not definite. Although 'non-immunologic' phagocytes, thyroid cells, amoebas, tumor cells and fibroblasts, ingest many of the same particles as polymorphonuclear leukocytes and macrophages, some studies using erythrocytes as test particles concluded that these phagocytes do not recognize 'immunologic' opsonins [160]. Other investigations, however, using bacteria or other particles, suggested that amoebas do respond to the opsonic effector molecules [161, 162].

The straightforward receptor model does not completely simplify understanding of a complex matter. Some reasons for this concern follow. As a first example, treatment of mouse peritoneal macrophages with either trypsin or chymotrypsin impairs their ability to ingest the protozoan parasite *Trypanosoma cruzi* [163]. Chymotrypsin does not keep these cells from ingesting C3-coated erythrocytes [163], which could mean that a *T. cruzi* receptor, discrete from C3, IgG and 'non-specific' receptors, exists. Indeed, following this logic, there could be separate receptors for all non-immunological debris that phagocytes take up. Second, whether or not a phagocyte ingests a given particle coated with an effector or opsonin depends on the state and type of the cell and on the amount and organization of the effector opsonin on the particle. Therefore, the number of possible interactions is considerable. Human monocytes, for instance, will bind but ingest poorly erythrocytes coated with IgG antibody directed against the D blood group [164, 165]. Normal human polymorphonuclear leukocytes neither bind nor ingest these erythrocytes [164, 165], but polymorphonuclear leukocytes taken from individuals with infections or myeloproliferative diseases both bind and ingest them [165]. Mouse peritoneal macrophages bind but do not ingest C3-coated erythrocytes, whereas 'stimulated' mouse peritoneal macrophages obtained from thioglycollate-induced exudates, bind and ingest erythrocytes opsonized with C3 [166]. All this information could mean that human monocytes and mouse peritoneal macrophages have 'uncoupled' IgG and C3 receptors, respectively, that normal human polymorphonuclear leukocytes lack IgG receptors, and that stimulated mouse peritoneal macrophages and polymorphonuclear leukocytes from sick humans have coupled their receptors. However, human monocytes and normal human polymorphonuclear leukocytes clearly ingest IgG-coated bacteria, platelets and other particles [118, 124, 145, 159, 167], and mouse peritoneal macrophages ingest particles which bind more opsonically-active C3 than erythrocytes [168]. Therefore, the variation in reagents brings out responses in the cells (for other examples, (see [118]), and without an exhaustive survey of conditions, firm generalizations about 'receptor' activity are hazardous.

Experiments purporting to show that 2-deoxyglucose inhibited ingestion of C3-coated and of IgG-coated particles but not the uptake of polystyrene particles were interpreted to mean that the mechanism coupling the C3 and IgG receptors but not the 'non-specific' receptors to the phagocytic effector system are energy-dependent [169]. Since the technique utilized to assay ingestion (microscopy) does not permit detection of even a large-scale inhibition of polystyrene particle ingestion [25]. these findings require corroboration by means of quantitative measurements of phagocytosis.

3.5.3 Binding

Experiments related to determining whether receptors are involved in binding of the effector to the cell began in 1927 when Bloom observed that oval-shaped antibody-coated avian erythrocytes clustered about rabbit pulmonary macrophages 'much in the manner of the petals of a daisy' [130]. Subsequently, numerous reports appeared concerning the binding of antibody- or complement-coated mammalian erythrocytes in 'rosettes' to all manner of phagocytes (reviewed in [118]), and since mammalian erythrocytes are round, the floral pattern invoked has changed to 'rosettes'.

The antibodies promoting binding of erythrocytes to phagocytes have, in almost all instances, been IgG molecules [164, 165, 171]. Mononuclear phagocytes yield rosettes with greater ease than polymorphonuclear leukocytes, although, depending on the conditions, erythrocytes will attach to neutrophils [172, 173].

Although many clear-cut parallels exist for the kinetics of ingestion and the phenomenon of rosette formation, in some respects the relationship is not entirely clear, nor is there good evidence that rosette formation is in any way comparable to the binding of soluble hormones to membrane receptors. Rosetting erythrocytes bound to monocytes have filopodial extensions interdigitating with similar structures on the phagocyte [174, 175]. These observations raise the issue as to whether rosetting is not simply a truncated version of ingestion itself.

An alternative approach to the kinetics of opsonin binding is represented by studies in which the attachment and ingestion of antibodies or else soluble or insoluble immune complexes to phagocytes have been investigated [20, 176–180]. The experimental systems utilized have permitted estimation of the number of binding sites and their binding affinities under specified conditions. On the one hand, immune complexes constitute a more manageable modality than solid particles for studying the interaction between activators of ingestion and phagocytes, and it will be of interest to see how kinetics of binding and ingestion compare in further studies. On the other hand, it remains to be proven that ingestion of soluble antibodies or microscopic immune complexes is really comparable to binding and ingestion of large particles. Some fraction of soluble complexes might be interiorized in the fluid phase of pinocytic vesicles coated with bound complexes, and the kinetics of such an interaction would differ from those of ingestion.

At the present time, ample evidence attests to the fact that certain molecules activate ingestion and, in some cases, promote binding of particles to phagocytes. But to ascribe the action of these effectors on phagocytes to receptor molecules is premature. The profound adhesivity of phagocytes for certain surfaces and for particles (and for each other) may be relatively non-specific and makes it improper to concoct a 'receptor' for every entity that binds to the surface of these cells.

Even with the confusion as to the nature of 'receptors,' some membrane molecules must receive the message, 'ingest' from appropriate particles, and glycoproteins or glycolipids are in the running as 'receptor' candidates. Proper clustering of these entities could be the initial step in recognition. Also consistent with the idea of mobile 'receptors' is the observation that human erythrocytes coated with anti-D IgG are not ingested but bind and migrate to one pole of monocytes or polymorphonuclear leukocytes, that is, they 'cap' [181]. Limited studies suggest that the movement of particles is energy-dependent, which is consistent with the concept that the receptors are linked physically to subplasmalemmal contractile elements.

Involvement of cell surface sugars in recognition is inferred from experiments in which the lectin concanavalin A inhibited [182] or enhanced phagocytosis of particles [183], induced pinocytosis [184], or opsonized erythrocytes [185]. Carbohydrate-containing blood group substances or mannose were reported to inhibit ingestion of other particles, possibly by binding to the particles, thereby blocking their interaction with cell surface glycolipids or glycoproteins [186, 187]. The finding that the presence or absence of defined glycoprotein blood group substances on erythrocytes determines whether the erythrocytes internalize malarial parasites [188] supports the idea that cell surface sugars are relevant to endocytosis.

3.6 RECOGNITION IN PINOCYTOSIS

The factors which 'induce' pinocytosis are no less complex than those that elicit phagocytosis and are less well studied. Agents which perturb cell surface sugars (lectins) [184] and lipids (lipo-proteins and lipid-soluble drugs) [189, 190] and a vast array of charged soluble substances all promote pinocytosis [191]. The common denominator, if any, is unknown. However, as suggested for phagocytosis, the ability of a given molecule to induce pinocytosis may be proportional to the number of membrane molecules (receptors) it can 'trigger'. Consistent with this view is the more avid pinocytosis by cells of aggregated rather than soluble substances and of multivalent in comparison with monomeric molecules [99, 176].

3.7 CONCLUSIONS

Concerning the molecular basis of endocytosis, little is certain save that phagocytosis is the result of a localized sequential recognition mechanism and is driven by the interaction of contractile proteins with each other and with the plasma membrane. The contractile process may govern or regulate the adhesion and fusion events of phagocytosis, as well as pseudopod assembly and movement. The precise nature of perturbations at the cell exterior that initiate these occurrences and the communication between recognition and the effector action still require explanation.

REFERENCES

1. Haeckel, E. (1862), *Die radiolaren,* Georg Reimer, Berlin.
2. Lewis, W.H. (1931), *Bull Johns Hopkins Hosp.,* **49**, 11–27.
3. Metschnikoff, E. (1887), *Ann. Inst. Pasteur,* **1**, 321–336.
4. Rodesch, F.R., Nève, P. and Dumont, J.E. (1970), *Exp. Cell Res.,* **60**, 354–360.
5. Kowalski, K., Babiarz, D., Sato, S. and Burke, G. (1972), *J. Clin. Invest.,* **51**, 2808–2819.
6. Cruickshank, C.N.D. and Harcourt, S.A. (1964), *J. Invest. Derm.,* **42**, 183–184.
7. Young, R.W. and Bok, D. (1969), *J. Cell Biol.,* **42**, 392–403.
8. Rabinovitch, M. (1968), *Semin Hematol.,* **5**, 134–155.
9. Spivak, J.L. (1973), *Scand. J. Haemat,* **11**, 253–256.

10. Abramson, N., Von Kapff, C. and Ginsburg, A.D. (1970), *N. Engl. J. Med.*, **283**, 248–250.
11. Garfield, R.E., Chacko, S, and Blose, S. (1975), *Lab. Invest.*, **33**, 418–427.
12. Parakkal, P., Pinto, J. and Hanifin, J.M. (1974), *J. Ultrastruct. Res.*, **48**, 216–226.
13. Fenn, W.O. (1921), *J. gen. Physiol.*, **3**, 575–593.
14. Altman, A. and Stossel, T.P. (1974), *Brit. J. Haemat.*, **27**, 241–246.
15. Gardner, D.E., Graham, J.A., Miller, E.J., Illing, J.W. and Coffin, D.L. (1973), *Appl. Microbol.*, **25**, 471–475.
16. Chambers, V.C. (1973), *J. Cell Biol.*, **57**, 874–878.
17. Roberts, J. and Quastel, J.H. (1963), *Biochem. J.*, **89**, 150–156.
18. Weisman, R.A., Korn, E.D. (1967), *Biochemistry*, **6**, 485–497.
19. Michell, R.H., Pancake, S.J., Noseworthy, J. and Karnovski, M.L. (1969), *J. Cell Biol.*, **40**, 216–224.
20. Ward, P.A. and Zvaifler, N.J. (1973), *J. Immunol.*, 111, 1771–1776.
21. Hållgren, R. and Stolenheim, G. (1976), *Immunology*, **30**, 755–762.
22. Mantovani, B., Rabinovitch, M. and Nussenzweig, V. (1972), *J. exp. Med.*, **135**, 780–792.
23, Stossel, T.P. (1973), *Blood*, **42**, 121–130.
24. Ulrich, F. and Zilversmit, D.B. (1970), *Am. J. Physiol.*, **218**, 1118–1127.
25. Stossel, T.P., Mason, R.J., Hartwig, J. and Vaughan, M. (1972), *J. clin. Invest.*, **51**, 615–624.
26. Ulrich, F. (1971), *Am. J. Physiol.*, **220**, 958–966.
27. Stossel, T.P., Alper, C.A. and Rosen, F.S. (1973), *J. exp. Med.*, **137**, 690–705.
28. Karnovsky, M.L., Shafer, A.W., Cagan, R.H. and Glass, E.A. (1966), *Trans. N.Y. Acad. Sci.*, **28**, 778–787.
29. Elsbach, P. (1972), *Semin. Hematol.*, **9**, 227–239.
30. Sastry, P.S. and Hokin, L.E. (1966), *J. Biol. Chem.*, **241**, 3351–3361.
31. Cline, M.J. (1966), *Nature*, **212**, 1431–1433.
32. Cohn, Z.A. (1966), *J. exp. Med.*, **124**, 557–571.
33. Goodall, R.J., Lai, Y.F. and Thompson, J.E. (1972), *J. Cell Sci.*, **11**, 569–579.
34. Ulsamer, A.G., Wright, P.L., Wetzel, M.G. and Korn, E.D. (1971), *J. Cell Biol.*, **51**, 193–215.
35. Sbarra, A.J. and Karnovsky, M.L. (1960), *J. biol. Chem.*, **235**, 2224–2229.
36. Karnovsky, M.L. and Wallach, D.H.F. (1961), *J. biol. Chem.*, **236**, 1895–1901.
37. Goodall, R.J. and Thompson, J.E. (1971), *Exp. Cell Res.*, **64**, 1–8.
38. Tizard, I.R. and Holmes, W.L. (1974), *J. Reticuloendothelial Soc.*, **15**, 132–138.
39. Lewis, W.H. (1939), *Z. exp. Zellforsch.*, **23**, 1–7.
40. Christiansen, R.G. and Marshall, J.M. (1965), *J. Cell Biol.*, **25**, 443–449.
41. Korn, E.D. and Weisman, R.A. (1961), *J. Cell Biol.*, **34**, 219–227.
42. Reaven, E.P. and Axline, S.G. (1973), *J. Cell Biol.*, **59**, 12–27.
43. Allison, A.C., Davies, P. and De Petris, S. (1971), *Nature New Biol.*, **232**, 153–155.
44. Wessels, N.K., Spooner, B.S., Ash, J.F., Bradley, M.O., Ludeña, M.A., Taylor, E.L., Wrenn, J.T. and Yamada, K.M. (1971), *Science*, **171**, 135–143.

45. Zigmond, S.H. and Hirsch, J.G. (1972), *Exp. Cell Res.*, **73**, 383–393.
46. Davis, A.T., Estensen, R. and Quie, P.G. (1970), *Proc. Soc. exp. Biol. Med.*, **137**, 161–164.
47. Klaus, G.G. (1973), *Exp. Cell Res.*, **79**, 73–78.
48. Malawista, S.E., Gee, J.B.L. and Bensch, K.G. (1971), *Yale J. Biol. Med.*, **44**, 286–300.
49. Axline, S.G., Reaven, E.P. (1974), *J. Cell Biol.*, **62**, 647–659.
50. Mimura, N. and Asano, A. (1976), *Nature*, **261**, 319–320.
51. Hatano, S. and Oosawa, F. (1966), *Biochim. biophys. Acta*, **127**, 488–498.
52. Adelman, M.R. and Taylor, E.W. (1969), *Biochemistry*, **12**, 4694–4975.
53. Weihing, R.R. (1976), *Biological Handbooks, Cell Biology*, (P.L. Altman and D.D. Katz, eds.), FASEB Publications, Bethesda, Md, **1**, 341–356.
54. Lehman, W. and Szent-Györgyi, A.G. (1975), *J. gen. Physiol.*, **66**, 1–30.
55. Shibata, H., Tatsumi, N., Tanaka, K., Okamura, Y. and Senda, N. (1972), *Biochim. biophys. Acta*, **256**, 565–576.
56. Bárány, M. (1967), *J. gen. Physiol.*, **50**, 197–216.
57. Pollard, T.D., Korn, E.D. (1973), *J. biol. Chem.*, **248**, 4691–4697.
58. Nachmias, V.T. (1974), *J. Cell Biol.*, **62**, 54–65.
59. Clarke, M. and Spudich, J.A. *J. mol. Biol.*, **86**, 209–222.
60. Stossel, T.P. and Hartwig, J.H. (1975), *J. biol. Chem.*, **250**, 5706–5712.
61. Adelstein, R.S. and Conti, M.A. (1975), *Nature*, **256**, 597–598.
62. Aksoy, M.O., Williams, D., Sharkey, E.M. and Hartshorne, D.J. (1976), *Biochem. biophys. Res. Commun.*, **69**, 35–41.
63. Pollard, T.D., Eisenberg, E., Korn, E.D. and Kielley, W.W. (1973), *Biochem. biophys. Res. Commun.*, **51**, 693–698.
64. Nachmias, V.T. and Asch, A. (1976), *Biochemistry*, **15**, 4273–4278.
65. Mockrin, S.C. and Spudich, J.A. (1976), *Proc. natn. Acad. Sci. U.S.A.*, **73**, 2321–2325.
66. Pollard, T.D. (1976), *J. Cell Biol.*, **68**, 579–601.
67. Pollard, T.D. and Korn, E.D. (1973), *J. biol. Chem.*, **248**, 448–450.
68. Stossel, T.P. and Hartwig, J.H. (1976), *J. Cell Biol.*, **68**, 602–619.
69. Clarke, M., Schatten, G., Mazia, D. and Spudich, J.A. (1975), *Proc. natn. Acad. Sci. U.S.A.*, **72**, 1758–1762.
70. Davies, W.A. and Stossel, T.P. (1976), *J. Cell. Biol.*, **70**, 296a,
71. Taylor, D.L., Rhodes, J.A. and Hammond, S.A. (1976), *J. Cell Biol.*, **70**, 112–143.
72. Pollard, T.D. and Ito, S. (1970), *J. Cell Biol.*, **46**, 267–289.
73. Pollard, T.D. and Korn, E.D. (1971), *J. Cell Biol.*, **48**, 216–219.
74. Boxer, L.A. and Stossel, T.P. (1976), *J. clin. Invest.*, **57**, 964–976.
75. Oosawa, F. and Kasai, M. (1971), In: *Subunits in Biological Systems*, Part A. (S.N. Timasheff and G.D. Fasman, eds.), Marcel Dekker Inc. New York, 261–322.
76. Hayashi, T. (1967), *J. gen. Physiol.*, **50**, 119–132.

77. Hartwig, J.H. and Stossel, T.P. (1975), *J. biol. Chem.*, **250**, 5699–5705.
78. Yagi, K., Mase, R., Sakakibara, I. and Asai, M. (1965), *J. biol. Chem.*, **240**, 2448–2454.
79. Ebashi, S. and Ebashi, F. (1965), *J. Biochem.*, **58**, 7–12.
80. Forer, A., Emmersen, J. and Behnke, O. (1972), *Science*, **172**, 774–776.
81. Spudich, J.A. (1972), *Cold Spring Harbor Symp. Quant. Biol.*, **37**, 585–593.
82. Spudich, J.A. and Lin, S. (1972), *Proc. natn. Acad. Sci. U.S.A.*, **69**, 442–446.
83. Löw, I., Dancker, P. and Wieland, T.H. (1975), *FEBS Letters*, **54**, 263–265.
84. Mason, R.J., Stossel, T.P. and Vaughan, M. (1973), *Biochim. biophys. Acta*, **304**, 864–870.
85. Kletzien, R.F., Perdue, J.F. and Springer, A. (1972), *J. biol. Chem.*, **247**, 2964–2966.
86. Hartwig, J.H. and Stossel, T.P. (1976), *J. Cell Biol.*, **71**, 295–303.
87. Weihing, R.R. (1976), *J. Cell Biol.*, **71**, 303–307.
88. Boxer, L.A., Hedley-Whyte, E.T. and Stossel, T.P. (1974), *N. Engl. J. Med.*, **291**, 1093–1099.
89. Berger, R.R. and Karnovsky, M.L. (1966), *Fed. Proc.*, **25**, 840–845.
90. Stossel, T.P., Pollard, T.D., Mason, R.J. and Vaughan, M. (1971), *J. clin. Invest.*, **50**, 1745–1757.
91. Fenn, W.O. (1923), *J. gen. Physiol.*, **5**, 143–167.
92. Garvin, J.E. (1961), *J. exp. Med.*, **114**, 51–73.
93. MacGregor R.R., Spagnuolo, P.J. and Lenter, A.L. (1974), *N. Engl. J. Med.*, **291**, 642–646.
94. McKeever, P.E. and Gee, J.B.L. (1975), *J. Reticuloendothelial Soc.*, **18**, 221–229.
95. Jones, B.M. (1966), *Nature*, **212**, 362–368.
96. Abercrombie, M., Heaysman, J.E.M. and Pegrum, S.M. (1971), *Exp. Cell Res.*, **67**, 359–367.
97. Helentjaris, T.G., Lombardi, P.S. and Glasgow, L.A. (1976), *J. Cell Biol.*, **69**, 407–417.
98. Weiss, L. (1971), *Fed. Proc.*, **30**, 1649–1657.
99. Cohn, Z.A. (1970), *Mononuclear Phagocytes*, (R. Van Furth, ed.), Blackwell, Oxford, pp. 121–128.
100. Casley-Smith, J.R. (1969), *J. Microscopy*, **90**, 15–30.
101. Wills, E.J., Davies, P., Allison, A.C. and Haswell, A.D. (1972), *Nature New Biol.*, **240**, 58–60.
102. Wagner, R., Rosenberg, M. and Estensen, R. (1971), *J. Cell Biol.*, **50**, 804–817.
103. Petitpierre-Gabathuler, M.P. and Ryser, H.P. (1975), *J. Cell Sci.*, **19**, 141–156.
104. Steinman, R.M., Silver, J.M. and Cohn, Z.A. (1974), *J. Cell Biol.*, **63**, 949–969.
105. Marchesi, V.T. and Steers, E. Jr., (1968), *Science*, **158**, 203–204.
106. Elgsaeter, A., Shotton, D.M. and Branton, D. (1976), *Biochim. biophys. Acta*, **426**, 101–122.

107. Penniston, J.T. and Green, D.E. (1968), *Archs. Biochem. Biophys.*, **128**, 339–350.
108. Schrier, S.L., Bensch, K.G., Johnson, M. and Junga, I. (1975), *J. clin. Invest.*, **56**, 8–22.
109. Zurier, R.B., Hoffstein, S. and Weissman, G. (1973), *Proc. natn. Acad. Sci., U.S.A.*, **70**, 844–848.
110. Davies, P., Allison, A.C. and Haswell, A.D. (1973), *Biochem. J.*, **134**, 33–41.
111. Hawkins, D. (1973), *J. Immunol.*, **110**, 294–296.
112. Henson, P.M. and Oades, Z.G. (1973), *J. Immunol.*, **110**, 290–293.
113. Johnson, R.B. and Lehmeyer, J.E., Jr., (1976), *J. clin. Invest.*, **57**, 836–841.
114. Goldstein, I.M., Horn, J.K., Kaplan, H.B. and Weissman, G. (1974), *Biochem. biophys. Res. Commun.*, **60**, 807–812.
115. Ignarro, L.J., Lint, T.F. and George, W.J. (1974), *J. exp. Med.*, **139**, 1395–1414.
116. Woodin, A.M. (1962), *Biochem. J.*, **82**, 9–15.
117. Woodin, A.M. (1975), *Exp. Cell Res.*, **92**, 201–210.
118. Stossel, T.P. (1975), *Semin. Hemat.*, **14**, 83–116.
119. Griffin, F.M. and Silverstein, S.C. (1974), *J. exp. Med.*, **139**, 323–336.
120. Griffin, F.M., Jr., Griffin, J.A., Leider, J.E. and Silverstein, S.C. (1975), *J. exp. Med.*, **142**, 1263–1282.
121. Griffin, F.M., Griffin, J.A. and Silverstein, S.C. (1976), *J. exp. Med.*, **144**, 788–809.
122. Hamburger, J.H. (1912, *Physikalisch-chemische Untersuchungen über Phagozyten,* J.F. Bergman, Munich, Wiesbaden.
123. Henon, M. and Delaunay, A. (1966), *Ann. Inst. Pasteur, Lille* **111**, 85–96.
124. Stossel, T.P. (1973), *J. Cell Biol.*, **58**, 346–356.
125. Debanne, M.T., Bell, R. and Dolovich, J. (1976), *Biochim. biophys. Acta,* **428**, 466–475.
126. Ignarro, L.J. and Cech, S.Y. (1976), *Proc. Soc. exp. Biol. Med.*, **151**, 448–452.
127. Sandler, J.A., Clyman, R.I., Manganeiello, V.C. and Vaughan, M. (1975), *J. clin. Invest.*, **55**, 431–435.
128. Schmidt-Gayk, N.E., Jakobs, K.M. and Hackenthal, E. (1975), *J. Reticuloendothelial Soc.*, **17**, 251–261.
129. Seyberth, H.W., Schmidt-Gayk, H., Jakobs, K.N., and Hackenthal, E. (1973), *J. Cell Biol.*, **57**, 567–571.
130. Manganiello, V., Evans, W.H., Stossel, T.P., Mason, R.J. and Vaughan, M. (1971), *J. clin. Invest.*, **50**, 2741–2744.
131. Masur, S.K., Holtzman, E., Schwartz, I.L. and Walter, R. (1971), *J. Cell Biol.*, **49**, 582–594.
132. Pearlman, D.S., Ward, P.A. and Becker, E.L. (1969), *J. exp. Med.*, **130**, 745–764.
133. Beachy, E.H. and Stollerman, G.H. (1971), *J. exp. Med.*, **134**, 351–365.
134. Fenn, W.O. (1921), *J. gen. Physiol.*, **4**, 373–385.

135. Mudd, S., McCutcheon, M. and Lucké, B. (1934). *Physiol. Rev.*, **14**, 210–275.
136. Van Oss, C.J. and Gillman, C.F. (1972), *J. Reticuloendothelial, Soc.*, **12**, 497–502.
137. Davies, W., Thomas, M., Linkson, P. and Penny, R. (1975), *J. Reticuloendothelial Soc.*, **18**, 136–148.
138. Wright, A.E. and Douglas, S.R. (1903), *Proc. Roy. Soc. Lond. Ser., B*, **72** 357–362.
139. Quie, P.G., Messner, R.P. and Williams, R.C. Jr., (1968), *J. exp. Med.*, **128**, 553–570.
140. Yasmeen, D., Ellerson, J.R., Dorrington, K.J. and Painter, R.H. (1976), *J. Immunol.*, **116**, 518–526.
141. Ciccimarro, F., Rosen, F.S. and Merler, E. (1975), *Proc. natn. Acad. Sci. U.S.A.*, **72**, 2081–2083.
142. Abramson, N., Gelfand, E.W., Jandl, J.H., Merler, E. and Rosen, F.S. (1970), *J. exp. Med.*, **130**, 1207–1215.
143. Okafor, G.O., Turner, M.W. and Hay, F.C. (1974), *Nature*, **248**, 228–290.
144. Gigli, I. and Nelson, R.A. Jr., (1968), *Exp. Cell Res.*, **51**, 45–47.
145. Johnston, R.B., Jr., Klemperer, M.R., Alper, C.A. and Rosen, F.S. (1969), *J. exp. Med.*, **129**, 1275–1290.
146. Smith, M.R., Shin, H.S. and Wood, W.B. Jr., (1969), *Proc. natn. Acad. Sci U.S.A.*, **63**, 1151–1156.
147. Jasin, H.E. (1972), *J. Immunol.*, **109**, 26–31.
148. Stossel, T.P., Field, R.J., Gitlin, J.D., Alper, C.A. and Rosen, F.S. (1975), *J. exp. Med.*, **141**, 1329–1347.
149, Wellek, B., Hahn, H. and Opferkuch, W. (1975), *J. Immunol.*, **114**, 1643–1645.
150. Mantovani, B. (1975), *J. Immunol.*, **115**, 15–17.
151. Bianco, C., Griffin, F.M., Jr., and Silverstein, S.C. (1975), *J. exp. Med.*, **141**, 1278–1290.
152. Scribner, D.J. and Fahrney, D. (1976), *J. Immunol.*, **116**, 892–897.
153. Huber, H., Polley, M.J., Linscott, W.D., Fudenberg, H.H. and Müller-Eberhard H.J. (1968), *Science*, **162**, 1281–1283.
154. Shurin, S.B. and Stossel, T.P. (1976), *Clin. Res.*, **24**, 450a.
155. Bjornson, A.B. and Michael, J.G. (1974), *J. Inf. Dis.*, **130**, Suppl. S127–S131.
156. Ward, P.A. and Zwaifler, N.J. (1973), *J. Immunol.*, **111**, 1777–1782.
157. Daughaday, C.C. and Douglas, S.D. (1976), *J. Reticuloendothelial Soc.*, **19**, 37–45.
158. Boxer, L.A. and Stossel, T.P., (1974), *J. clin. Invest.*, **53**, 1534–1545.
159. Rabinovitch, M., Manejias, R.D. and Nussenzweig, V. (1975), *J. exp. Med.*, **142**, 827–838.
160. Rabinovitch, M. (1970), *Mononuclear Phagocytes*, (R.Van Furth, ed.), Blackwell, Oxford, pp. 299–313.
161. Gerisch, G., Lüderitz, O. and Ruschmann, E. (1967), *Z. Naturforsch.*, **22b**, 109.

162. Stossel, T.P. Unpublished.
163. Nogueira, N. and Cohn, Z. (1976), *J. exp. Med.*, **143**, 1402–1420.
164. Abramson, N., LoBuglio, A.F., Jandl, J.H. and Rosen, F.S. (1970), *J. exp. Med.*, **132**, 1191–1206.
165. Zipursky, A. and Brown, E.J. (1974), *Blood*, **43**, 737–742.
166. Griffin, F.M., Jr., Bianco, C. and Silverstein, S.C. (1975), *J. exp. Med.*, **141**, 1269–1277.
167. Handin, R.I. and Stossel, T.P. (1974), *N. Engl. J. Med.*, **290**, 989–993.
168. Stossel, T.P. In: *The Granulocyte, Function and Utilization*, (G.A. Jamieson and R. Greenewalt, eds.), in press.
169. Michl, J., Ohlbaum, D.J. and Silverstein, S.C. (1976), *J. exp. Med.*, **144**, 1465–1483.
170. Bloom, W. (1927), *Archs. Path.*, **3**, 607–628.
171. Berken, A. and Benacerraf, B. (1966), *J. exp. Med.*, **123**, 119–144.
172. Messner, R.P. and Jelinek, J. (1970), *J. clin. Invest.*, **49**, 2165–2171.
173. Ross, G.D., Polley, M.J., Rabellino, E.M. and Grey, H.M. (1973), *J. exp. Med.*, **138**, 798–811.
174. LoBuglio, A.F., Cotran, R.S. and Jandl, J.H. (1967), *Science*, **158**, 1582–1585.
175. Douglas, S.D. and Huber, H. (1972), *Exp. Cell Res.*, **70**, 161–172.
176. Arend, W.P. and Mannik, M. (1973), *J. Immunol.*, **110**, 1455–1463.
177. Unkeless, J.C. and Eisen, H.N. (1975), *J. exp. Med.*, **142**, 1520–1532.
178. Phillips-Quagliata, J.M., Levine, B.B., Quagliata, F. and Uhr, J.W. (1971), *J. exp. Med.*, **130**, 589–601.
179. Steinman, R.M. and Cohn, Z.A. (1972), *J. Cell Biol.*, **55**, 616–634.
180. Mehl, T.D. and Lagunoff, D. (1975), *J. Reticuloendothelial Soc.*, **18**, 125–135.
181. Romans, D.G., Pinteric, L., Ralk, R.E. and Dorrington, K.J. (1976), *J. Immunol.*, **116**, 1473–1481.
182. Berlin, R.D. (1972), *Nature New Biol.*, **235**, 44–45.
183. Rabinovitch, M. and De Stefano, M.J. (1971), *Nature*, **234**, 414–415.
184. Edelson, P.J. and Cohn, Z.A. (1974), *J. exp. Med.*, **140**, 1364–1386.
185. Goldman, R. and Cooper, R.A. (1975), *Exp. Cell Res.*, **95**, 223–231.
186. Brown, R.C., Bass, H. and Coombs, J.P. (1975), *Nature*, **254**, 434–435.
187. Hokama, Y., Young, V. and Higaki, M. (1971), *J. Reticuloendothelial Soc.*, **9**, 451–464.
188. Miller, L.H., Mason, S.J., Dvorak, J.A., McGinniss, M.H. and Rothman, I.K. (1975), *Science*, **189**, 561–563.
189. Schwartz, S.L. (1976), *Fed. Proc.*, **35**, 85–88.
190. Yang, W.C.T., Strasser, F.F. and Pomerat, C.M. (1965), *Exp. Cell Res.*, **38**, 495–506.
191. Cohn, Z.A. and Parks, E. (1967), *J. exp. Med.*, **125**, 213–232.
192. Steinman, R.M., Brodie, S.E. and Cohn, Z.A. (1976), *J. Cell Biol.*, **68**, 665–687.

193. Stossel, T.P. (1976), *J. Reticuloendothelial, Soc.*, **19**, 237–245.
194. Boxer, L.A., Richardson, S. and Floyd, A. (1976), *Nature,* **263**, 259–261.

4

Virus Receptors

A. MEAGER and R.C. HUGHES
Division of Biological Sciences,
University of Warwick,
Coventry, U.K.,
and
The National Institute for Medical Research,
Mill Hill,
London, U.K.

Receptors and Recognition, Series A, Volume 4
Edited by P. Cuatrecasas and M.F. Greaves
Published in 1977 by Chapman and Hall, 11 New Fetter Lane, London EC4P 4EE

4.1 INTRODUCTION

Viruses, the simplest and smallest of life forms, are impotent, that is they lack the ability to reproduce outside a cellular organism. They have neither protein-synthesizing machinery nor energy supply. However, as we all know from the multitude of viral infections that afflict us and other creatures, viruses have been very successful in replicating themselves, and in making the necessary adaptations to survive the evolution of cellular organisms supporting their replication. Viruses are, of course, parasites that have acquired the universal knack of attaching to and penetrating living cells, both prokaryotes and eukaryotes, taking over the cellular processes for biosynthesis of proteins, glycoproteins, lipids and nucleic acids for the production of their own components, reassembling them into new virus particles (virions), and then re-emerging from the infected cells fully equipped to re-enact the infection process in neighbouring cells.

Virus infection, or the virus life-cycle, can be divided into six stages that follow chronologically:

(1) attachment or adsorption,
(2) penetration,
(3) uncoating,
(4) replication of nucleic acids and translation to produce viral proteins and glycoproteins,
(5) assembly of viral structural components, and
(6) release.

In this article we wish to concern ourselves with the first two stages in animal virus infection, with special emphasis on virus attachment. Other phenomena associated with virus attachment, such as haemagglutination, and fusion of viral and cellular membranes, will also be briefly discussed. We shall be concerned particularly with the surface of the virion, the cell surface and the molecular components that comprise these surfaces. It is known that, among individual viruses and host cells, these molecular components are extremely diverse. Thus, the complexity and diversity of the interacting components leading to virus attachment makes the virus-host cell system an excellent model for intercellular recognition and the specificity of cell surfaces. We shall describe in general terms the components required for virus attachment, and then, in the second part of the review, describe in detail individual virus-cell systems taken from the literature. In this way we shall build towards the idea of configurational complementarity between the virion surface and its cell surface receptor.

4.2 ANALYSIS OF THE COMPONENTS OF VIRUS ATTACHMENT

Virtually all our present knowledge about virus attachment comes from *in vitro* studies; the virus particles are suspended in a liquid medium overlying a monolayer of cells on a glass or plastic surface. Thus our current understanding of virus attachment is limited to the observations made in such systems and cannot therefore be extrapolated to all *in vivo* situations, although there may be superficial similarities. If this seems an unduly harsh opinion, it may be supported by the very few examples in the literature in which the mode of viral infection has been followed *in vivo* as well as in cultured cell lines. A recent series of studies [194–199] of nuclear polyhedrosis virus (NPV) replication in larval tissue of insects and in continuous cell lines, for example, shows that, contrary to the situation found in entry *in vivo* into columnar cells of the larval tissue, the entire virus particles are engulfed by cultured cell lines. *In vivo*, it appears that NPV virus is processed first at the surface of the columnar cells and only a portion of the virus, containing of course the viral nuclei acid, enters the cells. We shall discuss (Section 4.5.2) these alternative modes of entry of viruses into susceptible cells more fully. It is sufficient to point out here that the initial events in these two biologically different mechanisms for virus penetration into host cells need not be mediated via the same virus receptors at the host cell surface. If they are not, then we have to be clear which 'virus receptors' relate to different biological situations.

Although the attainment of optimal conditions for virus attachment in an artifically constructed virus-cell system may tell us little that is directly applicable to an *in vivo* situation, such a 'model system' should nevertheless greatly enhance our understanding of virus-cell interactions in the system *per se.* Therefore, we can proceed using such limited data, leaving the reader to speculate about the probably more complicated process of virus attachment occurring in nature.

4.2.1 Virus structure

Viruses, theoretically, can be built up in one of two ways – as a large complex single molecule or by multiple use of smaller molecules. Crick and Watson [1] suggested in 1956 that the protective coats of viruses are made up of identical protein molecules packed together in a regular manner. Thus, in many viruses, it has since been shown that there is only one protein in the viral coat, but with many copies of this protein in each particle. This is logical for viruses since, in general, they have limited amounts of genetic information in their nucleic acids and using one or a few proteins to form the viral coat or *capsid* makes the most economical use of it. The use of a large number of identical protein subunits means, however, that there are certain geometric limitations to the way in which viruses are put together.

The very simple viruses, which have only one coat protein can be rod-like, e.g. tobacco mosaic virus (TMV), or spherical (more correctly, icosahedral), e.g. turnip yellow mosaic virus (TYMV). The more complex animal viruses are generally spherical in outline, e.g. picornaviruses, with a few exceptions, e.g. vesicular stomatitis virus (VSV) which is bullet-shaped. In these more complex viruses, the nucleic acid or the *nucleo-capsid,* a coiled, TMV-like, rod-like structure of nucleic acid ensheathed in protein, is enclosed in the outer capsid, a protein or protein-lipid shell. The capsid is formed from one or more structural proteins arranged in a regular manner to produce an icosohedron in the non-lipid containing viruses. The electron microscope has revealed the virus capsid to be composed of morphologically identical, asymmetric units. These units, which are groups of structural proteins, are called *capsomeres.* The capsomeres have two faces, one being the face that forms the interior surface of the capsid, and the other, the exterior surface. It is this exterior surface in non-lipid containing viruses, e.g. poliovirus and reovirus, or certain domains on it that will form one of the colliding surfaces in virus attachment.

In more elaborate animal viruses, such as adenovirus, and in those viruses having a protein-phospholipid envelope, special viral proteins have been evolved which are directly involved in virus attachment. Adenovirus, a relatively large DNA-containing virus, has fibre-like protein appendages projecting from the apices of the capsid. Enveloped viruses, e.g. influenza, have glycoprotein 'spikes' extending outwards from the lipid layer. It is the surfaces at the ends of these projections or spikes that will take part in the collision with the cell surface.

To visualize what the exterior surface of virus capsids and projections look like we have to rely on images produced by electron microscopy. These electron-micrographs of virus particles usually show little detail of the structural complexity of the capsomere surface, primarily because of the small size of the virion. Nevertheless, electron micrographs of viruses and virus components are improving all the time, and the technique of freeze-etching is now yielding exciting results in this area, e.g. the adenovirus hexon capsomere [2].

At the molecular level we know that we are dealing with proteins and glycoproteins which make up the viral capsomeres and projections. These, of course, have three-dimensional structure, although, so far, there are not many examples where this has been elucidated in viral proteins. Viral structural proteins appear to be folded similarly to multimeric soluble proteins, with mainly hydrophilic charged amino acids on their outside facets, and neutral and hydrophobic amino acids in their interiors. The viral glycoproteins, which form the spikes of viral envelopes, have hydrophobic regions anchoring them into the phospholipid layer, with carbohydrate attached mainly to the regions of the molecules projecting out of this lipid layer. They are said to be 'amphipathic'. The carbohydrate moieties of the viral glycoproteins increase the hydrophilicity and probably the electrostatic charge of the spike regions. Thus, the surface of the virus, the capsomere or projection, will probably contain charged amino acid and sometimes carbohydrate residues. In the enveloped viruses, there is also a possibility that the polar regions of glycolipids which are composed of oligosaccharides will contribute to the charge and configuration of this interacting surface.

Summarising at this stage, we can say that much is known about the shape of viruses and their component capsomeres or envelopes [3–6, 156] although presently electron micrographs are of insufficient resolution for any great detail of surface structure to be apparent. X-ray crystallography has not been used much in the study of viral proteins, and hence detailed molecular structure of these is not available. Chemically, we know the nature of the components comprising the capsid or envelope surface (Table 4.1): these are proteins and possibly glycoproteins in non-envelope viruses, and glycoproteins and glycolipids in enveloped viruses. Little is known, however, about the detailed chemistry which underlies the molecular configuration(s) that will determine the stereospecificity of the collision or interaction with the host cell surface. Nor is it known to what degree movement is permissible between adjacent capsomeres or projections. Intuitively, one supposes that the glycoprotein spikes of enveloped viruses would be able to move more readily relative to each

Table 4.1 Enveloped and non-enveloped viruses and some representative members.

Viruses	Genera	Genome (mol. wt.)	Representative members
Non-enveloped viruses			
Picornaviruses	Enteroviruses Rhinoviruses Calciviruses	Single-stranded RNA	Polio, Coxsackie, Echo, EMC Foot and Mouth, Common cold
Reoviruses	–	Double-stranded RNA (segmented)	
Papovaviruses	Papillomoviruses Polyomaviruses	DNA (3×10^6)	Wart viruses, Polyoma, SV40
Adenoviruses	–	DNA (25×10^6)	Adenoviruses
Enveloped viruses			
Rhabdoviruses	–	Single-stranded RNA	Rabies, VSV, Potato yellow virus, Marburg (?)
Togaviruses (Arboviruses)	Alphaviruses Flaviruses	Single-stranded RNA	Sindbis, Semliki Forest virus, Japanese B encephalitis, Dengue
Myxoviruses	Ortho-(A, B and C)	Single-stranded RNA (segmented)	Influenza, Fowl plague
	Para-	Single-stranded RNA (single molecule)	Sendai, NDV, Measles
Oncornaviruses (RNA Tumour viruses)	Sarcoma Leukaemia	Single-stranded RNA (segmented or polyploid)	Rous, MSV MLV, Rausher, Moloney, Friend
Herpesviruses	–	DNA ($60–90 \times 10^6$)	Herpes, Epstein–Barr virus, Varicella
Poxviruses	–	DNA ($120–240 \times 10^6$)	Vaccinia, Variola (small pox)

other than the protein subunits of capsomeres in non-enveloped viruses simply because of the assumed fluidity of the underlying phospholipid layer in the former. Viral glycoproteins are closely packed and just how mobile they are in viral envelopes remains uncertain, however. In some cases the glycoproteins span the viral membrane to interact with internal proteins which may further restrict their movement [5, 6]. As a general picture, we can envisage the virus surface, the first component in virus attachment, as an electrostatically charged surface, often covered with projections, composed of proteins, glycoproteins and glycolipids or combinations of these, arranged in a regular manner and having varying degrees of freedom of movement relative to one another. It is perhaps best to think of the virus surface as being in a dynamic state, rather than the static 'fixed' entity depicted in electromicrographs, especially for enveloped viruses, since this dynamic state may figure importantly both in successful collision and in the subsequent fate of the virus.

4.2.2 Cell membrane structure

Cell membrane structure is itself the subject of several recent books and reviews [7–10]. We shall only deal with this briefly, since a full appreciation of the diversity and complexity of cell membranes is beyond the scope of our article.

There is still some controversy as to the state of the components comprising cell membranes, especially concerning the phospholipids of the 'lipid bilayer'. It is uncertain whether this phospholipid layer is crystalline and thus rigid, or fluid and thus mobile. The modern conception is that the phospholipid layer is largely fluid since experimentally it has been observed that proteins and glycoproteins are free to move or diffuse laterally within the lipid layer. The 'Fluid Mosaic Model' [11, 12] pictures a cell membrane as a 2-dimensional phospholipid liquid in which are embedded the various constituent proteins and glycoproteins, these being able to move laterally through this 'liquid'. The viral envelope may also be included in this model, but as pointed out previously the mobility of the glycoprotein projections may be more restricted because of close packing and anchoring to immobile internal proteins. In some cells too, such as erythrocytes (red blood cells), the membrane glycoproteins are restricted in their movement [13].

Many cell membrane proteins and glycoproteins intercalate the phospholipid layer and may extend right across this layer, and therefore they can be said to have amphipathic properties. In most cases, the hydrophobic regions of these proteins and glycoproteins lie buried in the hydrophobic interior of the phospholipid layer, whilst polar or hydrophilic regions, especially those carrying carbohydrate, extend outwards from the polar heads of the phospholipids. The very hydrophilic carbohydrate of glycolipids, which are integral components of membranes, will also face outwards into the external milieu.

In several instances, a subunit structure is evident in cell membranes, e.g. the red blood cell surface membrane [14]. However, in most of the cell membranes so far examined, subunit organization of the membrane components is not apparent and this may be due either to the preparative methods used to isolate membranes, which can destroy any regularity of structural configuration, or to the desiccating process used to fix membranes for electron microscopy, or to both. On another level, many cell membranes, particularly the plasma or surface membrane of tissue cells, exist as non-uniform structures and can be divided into several domains of different structure and function, e.g. the different 'faces' of liver cells (hepatocytes) [15]. There are numerous examples of viruses attaching *in vivo* to specialized regions of tissue cells, e.g. microvilli, suggesting that the distribution of 'virus receptors' *in vivo* may not be random, and thus confirming other evidence for the topological variation in surface structure of many cells.

Possibly the most revealing results in this area have been obtained using the freeze-etching technique with electron microscopy. The freeze-etch image is still a static one, but its 3-dimensional impression of the cell surface has made it possible to visualize several features, e.g. junctional complexes between cells, projections on the cell surface, and intramembranous particles [246].

Much of the remaining information in the literature about the cell surface has been obtained by the use of various molecular and viral 'probes'. These 'probes' all interact with the cell surface to elicit some change in organization or biochemistry which can be experimentally observed or estimated. For instance, lectins (plant proteins which bind specifically to certain carbohydrates), bind at the cell surface, especially to glycoproteins and glycolipids [16]. This lectin binding can in some cases lead to a reorganization of the glycoproteins within the surface membrane, and the phenomenon of 'patching' has been frequently observed [17, 18]. Enzyme treatment of the cell surface, especially by proteases and glycosidases, releases molecular and macromolecular components, and thus elicits changes in surface organization and properties. The various interactions of the cell surface with viruses can also lead to changes in structural organization and properties, e.g. virus-induced fusion, but these will be dealt with more fully at a later stage.

Thus there is a large amount of information already available to us concerning cell surface membranes, in particular an enormous literature on the individual components of these membranes. What is lacking, however, is the detailed chemistry of the various molecular and macromolecular components, either individually or, more frequently, in specific combinations, and how to visualize the membrane in a dynamic, rather than a static, state. Many cells, particularly those grown in culture, are growing and dividing, and therefore the surface membrane changes, both in size and composition, reflecting at each instant a point in the cell cycle. Cyclic variation to viral infection has indeed been found in some cases [225]. Over longer periods, cells, especially those of the developing embryo, differentiate, and again the surface membrane changes in composition, structure and properties. It should also not be forgotten that many well-differentiated cells and, to a certain extent, those cells grown under culture conditions, often secrete onto their surface, molecules and macromolecules which, in effect, coat this. This surface molecular coat, sometimes called the 'glycocalyx' [7–10], is composed largely of glycosaminoglycans and glycoproteins. There is some controversy as to whether these glycosaminoglycans and glycoproteins are 'extracellular' or should be regarded as part of the cell surface membrane. Nevertheless, it is obvious that it would be rash to propose any general picture of this second component of the collision between virus and cell. What we can say and be fairly sure of is that, besides the enormous diversity of animal cell types, there is also the dynamic nature of the cell surface to be considered. It is subject to continuous spatial, temporal and physicochemical changes and this has to be considered in the majority of cases of virus attachment. It is the dynamic character of the eukaryotic cell surface membranes which makes any mathematical treatment of virus attachment difficult [19]. (This contrasts with bacteriophage-bacterium system where the bacterial cell wall 'phage receptors' are fixed.).

4.2.3 Environmental factors

The third component involved in virus-host cell interactions is the medium that bathes the cells and carries the virus particles. In the case of physiological fluids the viscosity, temperature, constituents, and the movement of the fluid relative to the cells it bathes, vary considerably. The movement of the virus particles, and hence the rate of collision, will be governed by the nature of the fluid that carries them. Virus particles will often be in contact with molecules, particles and even cells, e.g. in the blood stream, which are carried by or themselves carry the virus particles. The nature of the physiological fluid will also greatly influence the structural configurations of the colliding virus and cell surfaces, e.g. the prevailing ionic conditions, pH and temperature, and the forces of attraction or repulsion between these.

The analysis of virus attachment, therefore, is a difficult task. It is a three component system, and not simply a two component, virus surface to cell surface system. All three components, the virus surface, the cell surface and the medium figure prominently in the successful collision between virus and cell. This collision only becomes a successful one when a bond of sufficient strength is formed such that the virus remains attached to the cell surface. This bond is apparently always a non-covalent bond, – a polar or ionic bond.

4.2.4 The virus receptor concept

So far we have not, in our conception of virus attachment, attempted to distinguish between 'loose' and 'tight' attachment (binding) nor commented on the reversibility of the process. However, virus particles can attach to many surfaces, even non-biological ones, e.g. glass, aluminium. The location or site of virus attachment will determine whether binding is loose or tight, reversible or irreversible, specific or non-specific, etc., and will determine subsequent events. For example, if a virus particle attaches to a glass surface it will remain there *ad infinitum,* the glass being inert. When, however, a virus particle attaches to a cell surface, or rather to discrete locations or sites on this surface, then the structural configuration and physicochemical properties inherent in these locations will determine whether the virus particle is bound reversibly or irreversibly, whether it will begin to penetrate the cell membrane or be engulfed by it, whether some enzymatic (viral or cellular) activity is elicited, whether membrane fusion occurs, and so on. All these changes are secondary to virus attachment, and are presumably specifically determined by the site or location of attachment.

The initial interaction between a virus particle and the host cell is likely to be dominated by electrostatic forces as these operate over relatively long distances [200]. Such forces are probably also involved in weak, non-specific interactions, e.g. with inert surfaces. The fact that most cell surfaces carry a net negative charge would form a barrier to those virus particles similarly charged. Nevertheless, despite the electrostatic barrier, attachment of such virions would eventually occur, the initial repulsive forces being overcome by the attractive forces operating when the cell surfaces approach closer than a critical separation (Curtis, 1973) and an interaction with a specific 'virus receptor' is allowed.

Virus receptors are those sites on the cell surface which recognize or are recognized by virus particles and provide specific points of entry into the cell. Of course, the probability of a cell generously providing receptors for viruses which inflict damage and death is unlikely. The existence of cell surface binding sites that serve both as phage or colicin receptors and as components of the uptake systems for low molecular weight compounds (vitamin B12) [254] and maltose [255] or iron [256–259] has been demonstrated in several bacterial systems. Thus phages ϕ80, ϕ80h, T5 and colicin M all appear to bind to a specific surface receptor for the siderophore, ferrichrome, and mutants lacking the receptor (*tonA*$^{-}$) are both phage-resistant and unable to utilize ferrichrome. In animal cells, too, it is the virus surface that has been adapted through evolution to attach to certain sites at the cell surface, and not the cell surface that has been adapted to receive the virus. 'Receptor' is perhaps the wrong word–*virus reception* (attachment) *site* is better since this also implies an object capable of being *recognized* (e.g. by a viral protein) and not itself being a *recognizer.* We shall use 'virus reception site' rather than 'virus receptor' in the remainder of this article.

Apart from purely electrostatic interactions, the attachment of viruses to the cell surface reception sites may involve protein–protein or protein–carbohydrate interactions (Fig. 4.1). In the latter case, either viral carbohydrate or host cell carbohydrate may play a role in attachment. It is well known, at least for the enveloped viruses, that viral carbohydrate is assembled by host cell enzymes and that the carbohydrate structure of viral and host cell glycoproteins and glycolipids are similar or identical [5, 6]. This raises the possibility that viral carbohydrate may be recognized as 'self' by host cells supporting viral replication. This is an intriguing possibility since carbohydrate specificity has often been suggested as underlying recognition between cells (see discussion in [201]. The viral surface may therefore mimic the host cell surface in similar interactions during infection. Unfortunately, at present the molecular details of virus attachment to cells are insufficient to decide if any of these interactions are important even in the model

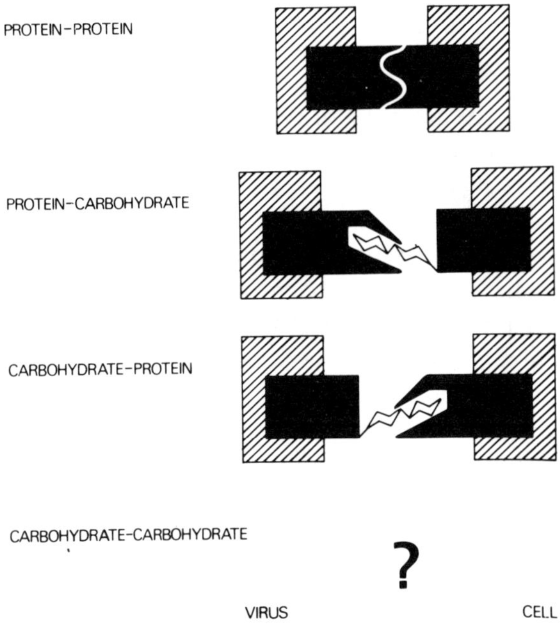

Fig. 4.1 Possible molecular interactions between viral and host cell surfaces. See text for discussion.

systems which have been studied most extensively. However, circumstantial evidence suggests that in different cases all of these may play a part in recognition between viruses and their host cells; although as mentioned previously other factors are likely to be involved in deciding whether or not a productive attachment occurs, allowing subsequent events in infection to be made.

4.3 NON-ENVELOPED VIRUSES

The non-enveloped viruses, all of which display cubic symmetry, can be, for our purposes, classified into two major groups; (a) those without any apparent specialized protein or glycoprotein projections, and (b) those with projections. Group (a) contains a very large number of viruses including *picornaviruses* (small RNA-containing virues), *reoviruses* (larger, double-stranded RNA-containing viruses) and *papovaviruses* (small DNA-containing viruses). Group (b) is largely restricted to *adenoviruses,* relatively large viruses having DNA as their genome.

4.3.1 Picornaviruses

Data concerning virus attachment is mostly available from *in vitro* studies involving picornaviruses. The picornaviruses are usually subdivided into three genera: (a) enterovirus, (b) rhinovirus, (c) calcivirus. Enteroviruses, so called because they initially cause infection in the alimentary tract, include poliovirus which is important from a disease standpoint in man. This genus also includes Coxsackieviruses (from the name of a town in the USA where the first outbreak was recorded), echoviruses, and a number of bovine, porcine, murine (e.g. EMC – encephalomyocarditis virus) avian and other viruses. Rhinoviruses are mainly viruses of man and horses – they cause the common cold in man. Foot and mouth disease (FMDV), which is a highly contagious, debilitating disease in cattle, is also a member of the rhinovirus genus. Not so much is known about the calciviruses which are mostly viruses of pigs and cats.

The picornaviruses have capsids of icosahedral symmetry with diameters ranging from 20–40 nm. All of the picornaviral structural proteins, of which there are four (60 copies of each), are present in equimolar amounts in the capsid. These four proteins are named VPI to VP4 in order of decreasing molecular weight [20]. The exact positioning of the viral proteins within the capsid is unknown, but some hypothetical models have been proposed [21, 22]. Nor is it known which of the four capsid proteins is outermost on the capsid surface; that is the protein which will first make contact with the virus reception site. VP4, the smallest protein, has been shown to be lost from the capside when, following attachment, a proportion of the virions spontaneously elute off and can no longer attach to cells [23, 24, 25], and it has been suggested that VP4 recognises the virus reception site [26] and remains attached to the cells during elution. However, other reports which have shown that eluted non-infectious virus is lighter in buoyant density than infectious virus, and thus may contain cell surface components including, perhaps, complete virus reception sites bound to the virus [27, 28] contradict this proposal. It is also difficult to reconcile the inaccessibility of VP4 to labelling with ^{3}H-acetic anhydride and lactoperoxidase-catalysed ^{125}I-iodination techniques, which label only exposed proteins, with its easy loss from virions and proposed role for virus attachment [57–59]. VP1 in native picornavirions appears to be the most heavily labelled protein when ^{3}H-acetic anhydride or ^{125}I-iodination are used to indicate externally located proteins in the picornaviral capsid [57, 58]. Thus, it is now suggested that VP4 alone cannot be responsible for virus attachment, and that the four picornaviral proteins act co-operatively to form the virion surface [29, 57], the whole of which, in native picornavirions, forms the viral recognition unit for attachment.

Many picornaviruses are limited to infection of the animal species which appears to be the natural host. Host range is now known, however, to depend on factors other than virus reception, although there are still a good many examples relating to picornaviruses where there is species specificity at the level of virus reception sites. Among these are the human picornaviruses which usually only attach to cells of primate origin [30, 31]. Some Coxsackieviruses can, however, infect mouse cells, and reception sites have been detected in the homogenates of the tissues of newborn mice, although they are decreased in number or absent in adult mouse tissues and in mouse L-cells [27, 32, 33, 34]. Other species differences regarding reception of Coxsackievirus subgroups, i.e. A and B, are given by Lonberg-Holm and Philipson [41]. Picornavirus reception also appears to be restricted to certain tissues within an individual host animal. For example, poliovirus reception sites are not found in heart or skeletal muscle tissues of primates [35, 36]. It is also possible to select from cultured cells, variant lines which lack reception sites, e.g. for poliovirus [37] and are resistant to virus infection. Hybrid cells between human and mouse cells have been constructed which in some lines contain poliovirus reception sites, and in others do not; loss of reception sites has been correlated with loss of human chromosome 19 from the human-mouse hybrids [38, 247, 249].

The virus reception sites for picornaviruses are present on cultured cells in only limited numbers ($1 \times 10^4 - 1 \times 10^5$), and can therefore be saturated with excess virus [24, 39–41, 60]. These reception sites make up only a small fraction ($\simeq/1\%$) of the total cell surface area, and so saturation does not result from spatial limitation [24]. Thus it should be possible to attach other viruses, assuming their reception sites are different, simultaneously or after attachment of one particular picornavirus, and indeed it has been shown that by binding firstly formolized type 1 polio virus to HeLa cell fragments neither the attachment of Coxsackie B viruses nor Newcastle disease virus (an enveloped virus of the paramyxovirus group) was affected [42]. In this study it was also demonstrated that other serotypes of poliovirus were prevented from attaching. Further work has confirmed this, and also shown that Coxsackievirus B1 and B3 compete for reception sites, but do not compete with the poliovirus serotypes for attachment [39]. From this and other more recent work, Philipson, Lonberg-Holm and co-workers [41, 43, 60] have proposed the concept of 'receptor families' – in our terminology, families of reception sites. Competition studies among several different picornaviruses indicated the presence of four different 'receptor families' on HeLa cells [43, 60]. Family one contains reception sites (receptors) for human rhinovirus types 2, 1A and 1B; family two, sites for human rhinovirus type 3, 5, 14, 15, 39, 41, 51 and Coxsackievirus type A21; family three, sites for the three poliovirus serotypes, and family four, sites for Coxsackie type B viruses. This is the first evidence that completely unrelated viruses attach to a common reception site. The reports [43, 60] also strongly suggest the existence of various configurations (at least four) of the viral capsid surface among picornaviruses which recognize the different families of reception sites.

The chemical nature of picornavirus reception sites has been difficult to determine, most probably because of the relatively low numbers of these per cell. Poliovirus reception sites on HeLa cells can be removed by trypsin, and Coxsackie type B virus sites by chymotrypsin [44, 45], and these sites were found to be regenerated fully in 1–2 hours and 2 days, respectively. These findings suggest a biochemical difference between the reception sites for poliovirus and Coxsackie type B virus. Poliovirus reception sites are also present in broken cells and plasma membrane fragments [31, 35, 46–50], but attempts to remove them from their *in situ* positions in the plasma membrane have been few. McClaren and co-workers [51] have reported the isolation and partial characterization of poliovirus reception sites after treatment of susceptible cells with 1% Triton X-100 and ultrasonic homogenization; the partly purified reception site complexes isolated from caesium chloride density gradients were found to contain glycoproteins and lipoproteins.

Further information relating to picornavirus attachment has been derived from studies involving either the chemical modification of the virus or modification of the cell surface. These indirect approaches have yielded some useful information as to the biochemical natures of the interacting viral and cell surfaces, and they have been fully reviewed by Lonberg-Holm and Philipson in their book [29, 41]. Most of the results reported in these studies concern haemagglutination of erythrocytes (see Section 4.5.1), but it is apparent that reagents which react with sulphydryl groups, e.g. para-hydroxymercuribenzoic acid (PHMB), 2, 3-dimercaptopropanol or oxidized British anti-lewisite (BAL), inactivate a number of haemagglutinating enteroviruses [52–55] and destroy poliovirus infectivity [56]. However, attachment of poliovirus to host cells does not require the presence of sulphydryl groups at the reception site [56].

Evidence that picornaviruses contain carbohydrate in their capsids which may participate in attachment is equivocal [41]. It would appear likely that some carbohydrates or carbohydrate-containing molecules can be adventitiously associated with picornavirions, and it is felt that as such they would play no major role in attachment.

Heat at 56–60 °C, ultrasonics and proteolytic enzymes (e.g. trypsin, chymotrypsin) denature picornavirus reception sites to varying degrees, and polyanions such as dextran sulphate, the lectin concanavalin A, and antibodies to cells (anticellular serum) interfere with picornavirus attachment [41, 60, 61, 62]. By contrast, neutralizing antibodies to picornavirions may not prevent attachment, e.g. poliovirus [63]. Lipoproteins are found in poliovirus reception site complexes [51], but lipase does not inactivate enterovirus reception sites [42, 46, 51, 64].

Other factors which affect picornavirus attachment are the ionic strength of the medium, divalent cations and, to a lesser extent, the pH of the medium [41]. Temperature probably also has an effect on picornavirus attachment [41, 60], and recently Noble-Harvey and Lonberg-Holm [65] have shown that HeLa cells have fewer reception sites for rhinovirus type 2 at 0°C than at physiological temperatures.

Summarizing here, we can deduce the following about picornavirus attachment: (a) there are relatively few picornavirus reception sites per cell, usually $10^4 - 10^5$; (b) attachment is specific; this specificity is reflected both in the viral capsid 'recognition unit' and in the virus reception sites at the cell surface; (c) the exact location of the viral 'recognition unit' on the capsid surface, its biochemical nature, and the composition of reception sites at the host cell surface are presently unknown; (d) attachment is subject to environmental factors. Thus, whilst the experimental evidence, so far, strongly suggests configurational complementarity between the picornaviral capsid recognition unit and the cellular reception site which governs the stereospecificity of the attachment, there is no biochemical confirmation of this.

4.3.2 Reoviruses

Rather less is known about attachment of reoviruses and papovaviruses, the other two types of virus in group (a). Reoviruses are structurally more complex than picornaviruses having distinct cores contained inside outer shells of capsomeres. The latter are composed of four proteins [66], one of which, designated μ_2, has been recently shown to be glycosylated [67]. Only a small percentage, 2–4%, of the μ_2 proteins contained in each reovirion have carbohydrate, and the significance of this finding in relation to attachment is currently unknown [67]. Reoviruses have a wide range of hosts and grow in many cell lines of different animal origin.

Most of the reports concerning reovirus attachment relate to erythocyte virus reception sites involved in haemagglutination [54, 71–77], and it is unknown whether these sites are the same biochemically as virus reception sites on susceptible host cells. Studies where the interaction of reoviruses [78–80] with susceptible host cells has been investigated contribute little to our understanding of early events in this process.

4.3.3 Papovaviruses

Like the reoviruses, papovaviruses are relatively complex virions containing several core proteins which appear to be histone-like and may be derived from the host cell. The papovaviral capsid is made up of two structural proteins, VP1 (75%) and VP2 (25%) [68, 69]. It is not known which of these proteins is located externally on the surface of the virion [70] and likely therefore to be responsible for recognition of host cell surface components [81, 82]. Recent work by Basilico and diMayorea [84] has underlined the fact that this specificity resides in the viral capsid. In this study, they obtained a polyoma virus mutant with altered viral coat proteins, and this showed a much reduced adsorption to (trypsinized) BHK and 3T3 cells compared with wild-type polyoma. Since freshly trypsinized cells in suspension were used for polyoma virus attachment, in this case presumably the specific attachment sites on the host cell surface are resistant to trypsin.

It is surprising that so little is known of the early interactions of papovaviruses with cell surfaces. A vast amount of information relating to the transformation of cells in culture by these tumour viruses has focussed attention on the cell surface membrane as that part of the cell which is most consistently altered during and after the transforming event. It is widely believed that these changes are similar to those which occur when cancer cells form *in vivo* from normal tissue cells. This subject has been recently reviewed [17, 18].

4.3.4 Adenoviruses

The group (b) of non-enveloped viruses includes only the adenoviruses, large DNA-containing viruses, which infect man, primates, cattle, pigs, mice, dogs and birds. There are many different serotypes, particularly of the human type, and they are responsible for minor respiratory infections, some kinds of pneumonia and conjunctivitis [85].

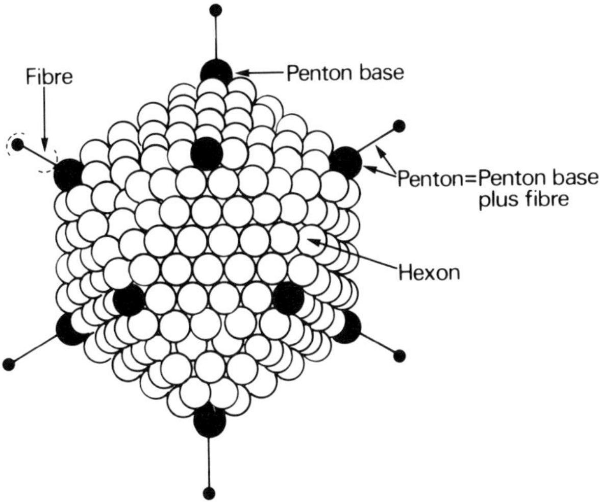

Fig. 4.2 The adenovirus showing the disposition of the major capsid proteins. Modified from Norrby [5] and Hughes [222].

The adenovirus capsid, of diameter 70–80 nm, has icosahedral symmetry, and 252 capsomeres (Fig. 4.2). Two hundred and forty of these capsomeres, called *hexons,* from the 20 triangular facets of a regular icosahedron. The remaining 12 capsomeres, called *pentons,* form the 12 apical structures [2, 85]. Pentons are composed of two different protein species: the penton base protein which fits directly into the holes at the icosahedral apices, and the fibre protein which projects outwards from each penton base protein [86, 87]. It is the fibre protein projections which, under the electron microscope, look like thin rods (2 x 9–31 nm) with knobs (4 nm diameter) at their external end [86, 88, 89], that are the adenoviral 'recognition units' for attachment [90, 91, 92]. The fibre protein is probably made up of three, possibly non-identical, polypeptide subunits, of molecular weight 60 000–65 000 [93, 94], some of which may have covalently linked carbodydrate residues [95]. Ishibashi and Maizel [95] found that radioactive glucosamine was incorporated into type 2 adenovirus fibre polypeptides, and estimated that there were about two residues of glucosamine, or its *N*-acetyl derivative, per polypeptide. This finding has not, however, been confirmed for type 5 adenovirus fibre protein [96, 97].

Many adenoviruses will infect cells other than those of their natural hosts. For example, most cultures of human cells are suitable for propagation of human adenoviruses, but productive or abortive infection also occurs in cultured rabbit, hamster, swine and monkey cells [85]. Abortive infection often leads to oncogenic transformation of the infected cells. Usually, an abortive infection is not due to a lack of adenovirus reception sites but in the case of type 5 adenovirus infection of mouse L-cells this is probably true [91]. An attempt to isolate mutants of BHK cells lacking reception sites for type 5 adenovirus was not successful [97].

There are approximately 10^4 adenovirus reception sites for types 2 and 5 per KB or HeLa cell [90, 91]. As mentioned earlier, it is the adenovirus fibre protein which is involved in attachment, and purified fibre protein can be bound to cells at the adenovirus reception sites. The number of fibre molecules binding to HeLa or KB cells at saturation was found to be 10 to 100-fold greater than the number of virions that could attach [90, 91]. This apparent discrepancy as to the number of reception sites can be explained either by steric limitations applying to virion attachment, or more likely, to the multivalent attachment of adenovirions, which each possess 12 fibre projections, to the reception sites. The adenovirus reception sites on HeLa cells are different from those of poliovirus [90], but type 2 adenovirus may share its reception sites, or some region of them, with Coxsackie virus type B3, a picornavirus [43]. Type 2 and type 5 adenoviruses probably share similar reception sites on HeLa cells [90], although competition experiments with purified type-specific fibre proteins suggests that the sites are non-identical (T. Butters and R.C. Hughes, unpublished results [222]). Similarly, the reception sites for adenoviruses type 2 and 3 are probably different [250].

The adenovirus reception sites on HeLa or KB plasma membranes can be removed by ionic or non-ionic detergent treatment [91, 98], sonic oscillation [98], and proteolytic enzyme digestion (pronase, subtilisin) [90]. However, digestion by trypsin or chymotrypsin, two other proteolytic enzymes, or by a mixture of neuraminidase and fucosidase, which remove cell surface sialic acid and fucose, apparently enhanced attachment of adenovirions [90, 99]. It is thought that such enhanced attachment could be brought about either because of decreased surface negative charge, or more likely, because of better exposure of reception sites, or a combination of both. The detergent-solubilized type 5 adenovirus reception site complexes are probably glycoprotein in nature [91]. Recent work by the authors, in which glycoprotein material extracted from KB cells with a fluorocarbon (Arklone P) into a water-soluble fraction was shown to bind specifically to type 5 adenovirus fibre protein immobilized on Sepharose, supports this view [96]. However, in this study it was not possible to determine from which part of the cell the glycoproteins binding to fibre were derived, although they were similar in molecular weights to authentic KB cell plasma membrane glycoproteins [100, 101]. The glycoproteins binding to fibre were carbohydrate-rich, but it is unknown whether fibre was bound through carbohydrate residues or amino acid elements of their polypeptide moieties [96].

There have been no studies relating to the chemical modification of the adenovirion or its cellular reception sites, except those of Neurath and co-workers on haemagglutination by type 7 adenovirus [102, 103]. It is unknown whether the adenovirus erythrocyte reception sites are chemically the same as reception sites on susceptible host cells.

Anticellular antibodies [104] and concanavalin A [91] inhibit attachment of adenoviruses to host cells. However, neutralizing antibodies to adenovirions, whilst causing aggregation and destroying infectivity, probably do not completely prevent attachment since the toxic effect of adenovirions mediated by pentons [105, 106] is unaffected or increased [107, 108]. It is assumed that penton toxicity first requires attachment of the fibre projections with their specific attachment sites on the cell surface. Heat at 56 °C causes a morphological change in adenovirions; the penton subunits with fibre proteins and associated peripentonal hexons are lost [92, 109]. Such heat-treated adenovirions are, of course, non-infectious since they have lost their vehicles of attachment.

In conclusion it can be said that many features of adenovirus attachment to host cells have been characterized. The adenovirus recognition unit on the capsid surface is clearly the fibre protein projection. This fibre protein which is a complex of three polypeptides of approximately equal molecular weight that possibly are glycosylated [93–95], is tubular and has a terminal 'knob' which must be the first structural element to make contact with the cell surface and the reception sites. Once the virus has become attached to one reception site through an apical fibre protein, some of the other remaining fibre proteins could attach to adjacent reception sites to give firmer anchorage of the adenovirion and perhaps facilitate entry into the cell. The host cell reception sites probably contain glycoproteins, but it is as yet unknown whether fibre protein attaches to these by way of their carbohydrate residues or by their amino acids. The fact that adenoviruses attach to a large variety of different animal cells suggests that either the stereospecificity of the interaction between virion and cell is not strict, or that the biochemical components of the reception sites are common to many cell types.

4.4 ENVELOPED VIRUSES

The second major category of animal viruses have glycoprotein–phospholipid envelopes which encapsulate the viral nucleic acid, nucleocapsid or internal capsid, e.g. influenza virus and herpes simplex virus. For our purposes the components of the external surface of the viral envelope are of prime importance since these are concerned in attachment to susceptible host cells. The envelopes of all the viruses we shall be concerned with are grossly similar. They are all composed of a phospholipid bilayer derived from the infected host cell plsma membrane into which a large number of copies of one or more viral glycoproteins are inserted. These glycoproteins are packed together in a regular manner to form 'spikes'. The viral envelopes also contain glycolipids which may conttribute to the conformation of the envelope surface. In general, under the electron microscopic, no region of the envelope surface appears distinctly different from any other region, and it can probably be assumed that any one glycoprotein or glycoprotein-glycolipid spike complex will act as a recognition unit for attachment to host cells.

As mentioned previously, it is unlikely that the genetic content of most viruses is sufficient to specify a large number of enzymes manufacturing virus-specific membrane lipids. In general, the lipids are synthesized by enzymes coded for in the host cell. Similarly the synthesis of a typical viral envelope glycoprotein, for example of Sindbis virus, would require at least five sugar transferases for assembly of the carbohydrate units and one or two enzymes for synthesis of the attachment site to polypeptide. It seems likely, therefore, that the carbohydrate structure is largely, although not necessarily entirely, specified by host cell glycosyl transferases acting on the virally coded polypeptide. Presumably these polypeptides contain a suitable amino acid sequence recognized by a host cell glycosyl transferase [206] and enabling the first sugar to be linked to the viral protein. Once this linkage is formed, the chain may grow by the concerted action of other host cell glycosyl transferases. It seems probable that this host cell modification of a virally coded protein accounts for certain antigenic similarities between the membranes of many viruses and the host cells in which they grow [203–205]. It may, therefore, be important to consider, in the following sections, the different genetic origins of polypeptide and carbohydrate components of the viral envelope glycoproteins and their relation to specific virus–host cell interactions.

4.4.1 Rhabdoviruses

Rhabdoviruses include rabies virus, which is topical at the moment, and vesicular stomatitis virus (VSV) which causes an infection in cattle very similar to foot and mouth disease. Morphologically, the rhabdovirions appear bullet-shaped in the electron microscope. The envelope is studded with projections (spikes) [110], and these are the rhabdovirion glycoproteins [111]. The rhabdoviruses are the simplest viruses in the group from the viral envelope point of view since they have but one glycoprotein species. There are one or two other viral proteins in the envelope, but these are non-glycosylated and probably form the inner surface of the envelope which surrounds the ribonucleoprotein core; they are called M proteins or matrix proteins [111].

The rhabdovirus glycoprotein, especially that of VSV, has been well-characterized [112–114]. The VSV glycoprotein has a molecular weight of about 67 000 [115], and contains 9–10% by weight carbohydrate [112, 116]. This glycoprotein contains mannose, galactose, *N*-acetylglucosamine and neuraminic acid as the major carbohydrate components, and *N*-acetylgalactosamine and fucose as more minor components. These sugars are contained in two to four oligosaccharide chains which appear to be heterogeneous both with respect to sequence and size [112, 116–118]. There are quantitative differences in carbohydrate composition when VSV is grown in different mammalian cell lines [116–118] and neuraminic (sialic) acid is absent if VSV is grown in mosquito cells [119, 245]. These differences reflect the different levels of sugar (glycosyl) transferase activities present in the various infected host cells.

Moyer and co-workers [118] have proposed the following sequence:

$$\text{(Man)n} \rightarrow \text{GlcNAc} \rightarrow \overset{\overset{\text{Fuc}}{\downarrow}}{\text{GlcNac}} \rightarrow \text{Asn}$$

as the carbohydrate-peptide linkage region of the major class of oligosaccharides of VSV glycoprotein of virus grown in HeLa cells.

The VSV glycoprotein, or more correctly its carbohydrate-containing spike region, certainly acts as the viral recognition unit for attachment. Proteolytic enzyme (e.g. trypsin) treatment of purified VSV virions removes the glycoprotein spike regions [121–123] leaving a lipophilic terminal peptide fragment associated with the envelope lipid bilayer [123], and the resulting spikeless particles are non-infectious. Infectivity can be restored to spikeless VSV particles by the specific addition of purified VSV glycoprotein [124]. Thus the intact VSV glycoprotein is required for attachment but, more interestingly, Schloemer and Wagner [119, 121, 125] have shown that neuraminic acid attached to the glycoprotein oligosaccharide moieties is critically involved in attachment. Schloemer and Wagner found that neuraminidase-treated VSV or VSV grown in mosquito cells which lacks glycoprotein neuraminic acid are only weakly infectious and fail to attach well to host cells. In both cases, resialylation restored or enhanced attachment and hence infectivity [119, 121]. VSV envelope surface sialoglycolipids, which most probably are integrated exclusively into the outer lamella of the lipid bilayer [126] were found not to contribute to VSV infectivity [121], showing that neuraminic acid must be present on a glycoprotein for recognition by the host cell surface. When VSV is grown in a mutant line of Chinese hamster ovary (CHO) cells which lack an *N*-acetylglucosaminyl transferase the VSV glycoprotein is under-glycosylated [127]. However, the VSV grown in such cells is as infectious as that grown in wild-type CHO cells, and it is suggested by Schlesinger and co-workers that synthesis of heterosaccharides with those neuraminic acid residues involved in attachment is unaffected by the mutation [127]. Recent work in our laboratory with a similar mutant of BHK cells in collaboration with Dr S. Rosen [139, 140] provides direct support for this proposal. Although some of the carbohydrate chains of mutant surface glycoproteins appear to be incompletely glycosylated and no longer bind to the toxic lectin ricin (consequently these cells are very resistant to ricin-induced cell killing), the glycoproteins bind to ricin (and the cells revert to a sensitive state) after treatment with neuraminidase. One plausible interpretation of these findings is that new ricin-binding sites are revealed on surface glycoproteins of the mutant cells by removal of sialic acid. Presumably, sialic acid groups present in these carbohydrate chains prevent ricin binding while, in the case of the carbohydrate chains which, in the mutant cells, are not completed due to the enzyme deficiency, sialic acid groups do not interfere with ricin binding. It is probable therefore that the VSV glycoproteins made in mutant cells have incomplete carbohydrate chains, but those chains bearing sialic acid vital for attachment are unmodified.

While it now seems well established that the viral recognition unit for VSV involves neuraminic acid on the VSV glycoprotein [248], the biochemical nature of the virus reception site on host cells is not so clear. There are three main possibilities: (a) the viral neuraminic acid residues may exert an effect by modulation of the surface charge on the cell surface. Although the surface of most animal cells carries a net negative charge, a sialic acid-rich viral surface may make contact at patches of the host cell surface which are cleared of acidic host cell surface components, mainly sialoglycoproteins [253]. However, the fact that sialic acid-containing glycolipids do not substitute for viral sialoglycoproteins in restoring infectivity to neuraminidase-treated virions argues against this interpretation; (b) the host cell surface may contain a protein (or glycoprotein) which combines with neuraminic acid, or an oligosaccharide sequence containing neuraminic acid, in the viral glycoprotein; (c) the host cell protein (glycoprotein) may combine with a complementary peptide sequence in the viral glycoprotein, the configuration of which may require a complete content of neuraminic acid. This last model has been proposed to account for the absolute requirement for sialic acid in the expression of the human blood group MN antigenic determinants [10, 202]. As VSV will infect a very large number of different animal cells, including invertebrate cells, it is probable that the reception sites are common constituents of cell membranes. Schloemer and Wagner [125] have shown that trypsin and neuraminidase treatment of L-cells, CHO and Madin–Darby bovine kidney (MDBK) cells results in enhanced attachment of VSV. Therefore, neuraminic acid at the cell surface cannot be involved in attachment and its removal may even enhance attachment, since trypsin digests some of the surface glycoproteins and neuraminidase removes surface neuraminic acid, thus reducing the net negative charge on the cell surface. Alternatively, the VSV reception sites may be better exposed after enzyme treatment.

4.4.2 Arboviruses (Togaviruses)

This group of enveloped viruses is further divided into Alphaviruses and Flaviviruses [141]. With the exception of Japanese B encephalitis virus, a flavivirus, all studies involving attachment of this group of viruses to cells have been done with alphaviruses. Alphaviruses are positive-strand RNA viruses, with diameter 40–80 nm, and consist of a nucleocapsid with

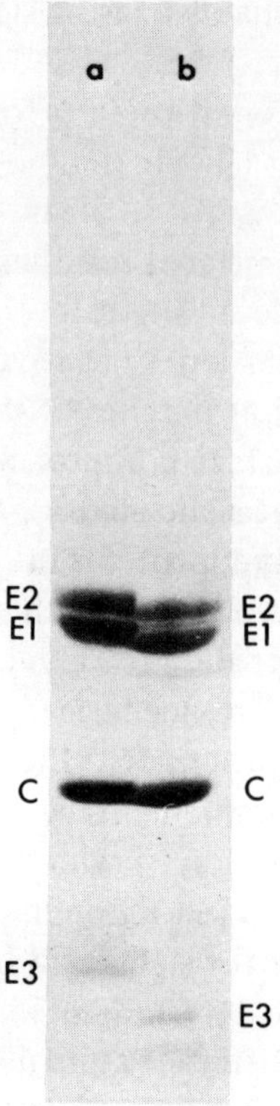

Fig. 4.3 Glycoproteins of Semliki forest virus grown (a) in BHK cells and (b) in an enzyme-deficient BHK cell mutant. The greater mobility of the glycoproteins of virus produced in the latter cells is attributed to a lower amount of carbohydrate. The properties of the under-glycosylated virions are discussed in the text. Photograph kindly supplied by Dr I. Kennedy.

cubic symmetry, and an envelope studded with glycoprotein projections [85]. They include various encephalomyelitis viruses, Sindbis virus and Semliki Forest virus; the latter two viruses are the most studied in the group.

Alphaviruses have two or three glycoproteins in their envelopes, forming the projections or spikes. In both Sindbis and Semliki Forest (SFV) viruses there are two 'large' glycoproteins, called E1 and E2, of apparent molecular weights 49 000–54 000 [128–130], and in SFV a small glycopeptide, E3, of apparent molecular weight 10 000 is also present [131, 132] (Fig. 4.3). E1 contains 7.5%, E2 11.5% and E3 as much as 45% carbohydrate by weight in SFV [132] and E1, E2 of Sindbis virus each contain about 8% carbohydrate by weight [133]. The monosaccharides include *N*-acetylglucosamine, mannose, galactose, fucose, and sialic acids in both SFV [132] and Sindbis [133], and are present in both A-type and B-type oligosaccharides [134]. The exact composition and size of these oligosaccharides, however, is probably determined by the host cells in which the viruses are grown [135, 136].

The glycoprotein projections of alphaviruses, e.g. SFV, are necessary for virus infectivity [127, 137, 138]. Thermolysin removes the envelope spikes leaving small lipophilic polypeptide moieties of E1 and E2 inserted in the lipid bilayer [137]. The resulting spikeless particles were weakly infectious probably because they aggregate badly and do not attach to cells [137]. However, in contrast to VSV, the terminal neuraminic acid residues on E1 and E2 (also E3) in both SFV and Sindbis virus do not have any role in attachment [136, 138]. Neuraminidase-treated SFV [138] or Sindbis virus grown in mosquito cells [136], both of which contain no neuraminic acid, have an infectivity equal to their respective control viruses. The growth of Sindbis virus in mutant CHO cells deficient in an *N*-acetylglucosaminyl transferase, which results in under-glycosylation of E1 and E2, has no effect on progeny virus infectivity [127]. Similar findings have been made with SFV grown in a mutant BHK cell line, Ric^R14 [139, 140], also lacking *N*-acetylglucosaminyl transferase activity (A. Meager, S.I.T. Kennedy and R.C. Hughes, unpublished results). In this case, all three envelope glycoproteins, especially E3, have reduced amounts of carbohydrate (Fig. 4.3); the monosaccharides which appear to be reduced in number are mannose, galactose, *N*-acetylglucosamine, and sialic acid. Extensive removal of protein-bound carbohydrate [138] by treatment with a mixture of glycosidases did, however, lead to loss of infectivity, but since the virions aggregated when stripped of carbohydrate it could not be ascertained whether loss of infectivity was simply due to non-attachment of virions. Probably a full complement of carbohydrate in the viral glycoproteins is required in attachment of Sindbis virus to some cells. Thus, Sindbis virus grown in mutant CHO cells [127] fails to haemagglutinate goose erythrocytes.

There appear to be approximately 10^5 reception sites on the cell surfaces of chicken or BHK cells for Sindbis virus [142], and these tend to cluster on virus attachment. This suggests that the reception sites are able to move laterally through the surface membrane. The biochemical nature of the reception sites in unknown, although Mooney *et al.* [143] have shown that Sindbis will attach to liposomes, artificially constructed protein-free phospholipid membranes, made from sheep erythrocyte extracts. They suggest that initially Sindbis virus attachment only requires a lipid membrane containing phospholipid (phosphatidyl ethanolamine) and cholesterol, and that neither glycoproteins nor glycolipids are involved in the attachment step. While this may explain why Sindbis virus can attach to the cell surfaces of many different cell types, phospholipid and cholesterol being ubiquitous, it does not explain the discreteness of the Sindbis virus reception sites observed by Birdwell and Strauss [142] nor their ability to cluster.

That a higher degree of reception site specificity is required for attachment of arboviruses rests on studies involving the flavivirus, Japanese B encephalitis virus. Nozima and co-workers [144–148] have shown that *d*-mannose and the polysaccharide mannan inhibited haemagglutination by Japanese encephalitis virus, and α-mannosidase destroyed the activity of purified erythrocyte and host cell reception sites. Thus, there appears to be clear evidence that mannose is required for attachment at the cellular reception sites. Whether mannose residues, probably protein-bound, provide reception sites for all flaviviruses, and other related viruses, on susceptible host cells remains to be tested. Mannose certainly has a wide enough distribution (it is present in all mammalian cell surface membranes) for it to function as such. Indeed, the assumption of such a role is believed to be an important factor in bacterial invasiveness, e.g. attachment of enterobacteriacae to digestive tract mucosal cells [251]. So far, however, there is little direct evidence for a viral envelope protein or glycoprotein with lectin-like activity that could bind to mannose residues of host cell reception site glycoprotein(s).

4.4.3 Myxoviruses

This group contains the viruses responsible for influenza and a number of other respiratory illnesses. Myxoviruses are really two groups of quite distinct, serologically unrelated viruses, and these are now known as orthomyxoviruses and paramyxoviruses. They were originally called myxoviruses by Andrewes *et al.* [149] because of their affinity for certain mucoproteins and because they contain an enzyme (neuraminidase) which removes neuraminic acid from mucoproteins (glycoproteins). Subsequently, Waterson [150] subdivided the group into orthomyxoviruses (80–120 nm diameter) and paramyxoviruses (100–200 nm) largely on the basis of size. Although the two subgroups are morphologically similar externally, it is now recognized that there are substantial biochemical differences between them. For example, the orthomyxovirus RNA genome is segmented, while the paramyxovirus RNA genome is one unbroken strand [85, 151]. The haemagglutinin and neuraminidase activities of orthomyxoviruses are present in two distinct glycoproteins [151a), whilst these activities appear together in one paramyxovirus glycoprotein [151b, 152]. The haemagglutinin glycoprotein of

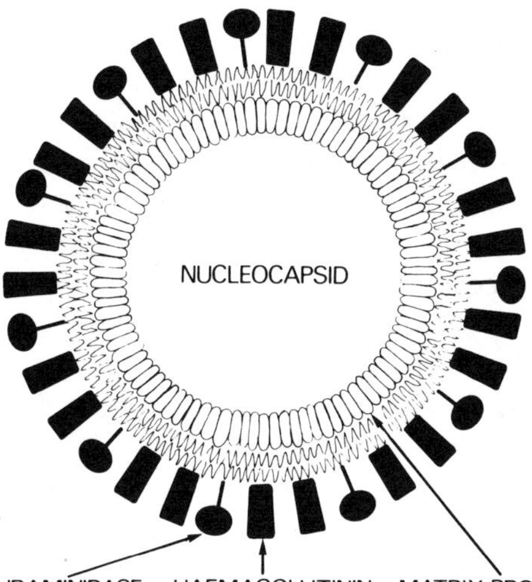

Fig. 4.4 Model structure of the myxovirus adapted from Lenard and Compans [156].

orthomyxoviruses may be cleaved into two polypeptide chains, heavy (mol. wt. 70 000) and light (mol. wt. 50 000) in mature virions, whereas the paramyxovirus haemagglutinin-neuraminidase glycoprotein is one uncleaved polypeptide [151b].

In the influenza virus (orthomyxovirus) envelope (Fig. 4.4) the two glycoproteins, haemagglutinin and the neuraminidase, form separate projections [151a]. Both haemagglutinin, which appears to be trimeric (3 protein subunits), and neuraminidase, which is probably tetrameric (4 protein subunits) project from the virion to the same extent [151a,d], and hence both could take part in attachment to host cells. There are about 3–6 times more haemagglutinin (HA) projections (approximately 500–600 per virion) than neuraminidase projections per virion [151a]. If each spike is considered to have a radius 2.5 nm [207, 208] and the spikes are separated by about 7–7.5 nm [209] it can be calculated that at least one quarter of the viral surface is occupied by spikes.

Antibodies to neuraminidase do not prevent attachment [151d, 153, 154]. Thus it is felt that the HA projections are the viral components which are required for attachment. Stronger evidence that the neuraminidase projections play no part in attachment is provided by the finding that mild trypsin treatment of virions specifically removes the neuraminidase from the viral envelope without affecting infectivity

[151d]. Removal of the HA, however, destroys infectivity [151d]. The biological role of viral neuraminidase is still problematic. Early studies [210] using genetic variants of influenza virus having a functionally altered neuraminidase of low activity, suggested that newly synthesized virions were released from the host cell membrane very slowly during viral replication. Similarly, antibodies directed against the enzyme completely suppressed the release of virus without any appreciable effect on synthesis of viral components [211–214]. In an electron microscope study [216] carried out with cells infected with influenza virus and incubated with anti-neuraminidase serum, a marked collection of viral particles near the cell surface was observed, as though some stage in viral release was blocked by the antibody. However, the situation is clearly complex since a monovalent anti-neuraminidase which completely blocked catalytic activity had no effect on viral release when added to cultures of infected cells [215]. Presumably, the inhibitory action of divalent anti-neuraminidase antibody on viral release cannot be related to a direct blocking of neuraminidase activity and thus an effect on elution of the newly synthesized virions. Perhaps the effect of the divalent antibody on viral replication lies in its ability to freeze areas of the plasma membrane of the host cell containing viral glycoproteins, including the neuraminidase. These molecules are inserted into the surface of host cells supporting the lytic infection, at the budding sites through which the newly assembled nucleocapsids bud on their exit from the cell. It is conceivable that cross-linking of glycoproteins in these budding sites by divalent antibodies may decrease the flexibility of membrane structure, necessary for bud formation.

In the case of paramyxovirions, such as Sendai virus and Newcastle disease virus, where the glycoprotein projections contain both HA and neuraminidase in a single glycoprotein species [151b, 152], possibly some defined region in the glycoprotein complex separate from the active site of neuraminidase is the viral recognition unit. However, in these viruses the picture is more complex since they possess a second, smaller glycoprotein species (mol. wt. 53 000) which is probably responsible for the haemolysing and cell fusing activities found in paramyxoviruses [151b]. It is presently unknown whether this glycoprotein forms separate projections from the HA-neuraminidase glycoprotein (i.e. similar to the separate projections for HA and neuraminidase glycoproteins in influenza [151a]) or whether they form a complex projection with the HA-neuraminidase glycoprotein.

Relatively little work has been done on the carbohydrate structure of myxovirus glycoproteins. Simian Virus 5 (SV5), a paramyxovirus, has glycoproteins containing galactose, mannose, *N*-acetylglucosamine and fucose [155], and these common monosaccharides are probably present in the glycoproteins of most other myxoviruses. Neuraminic acid is, however, absent from viral glycoproteins and glycolipids in myxoviruses possessing neuraminidase [156] (N.B. not all paramyxoviruses contain neuraminidase, e.g. measles virus). Unlike VSV, the absence of terminal sialic acid residues does not affect myxovirus attachment to host cells [151d]. Furthermore, if influenza virions are sialylated such that the HA glycoproteins gain terminal sialic acid residues, infectivity is apparently unaffected or even increased for some cell lines. The haemagglutinating activity is almost abolished, however [151d]. Schulze [151d] suggests that this means either the host cell and erythrocyte reception sites are different, or more probably that the sialylated virions can no longer form virus-erythrocyte lattices (see Section 4.5.1).

There is good evidence that the myxovirus glycoprotein projections recognize terminal sialic acid residues on cell surfaces [157, 158, 159]. The sialic acids can be present in glycoprotein oligosaccharides or in sialoglycolipids (gangliosides) [160, 161]. Probably the best illustration that sialic acid acts as the reception site for myxoviruses comes from work involving the construction of artificial membranes or liposomes. Tiffany and Blough [160] showed that myxoviruses, the PR8 strain of influenza and the Sato strain of Newcastle disease virus, attached to an artificial lipoprotein membrane consisting of phosphatidyl-choline and the glycoprotein fetuin which contains sialic acid. Neither type 2 nor type 5 adenoviruses nor herpes virus could attach to the artificial membrane containing fetuin [160], and so presumably fetuin has no reception sites for these viruses. Recently, Haywood [161] has shown that gangliosides containing sialic acid, when incorporated into liposomes, will attach Sendai virus. When the ganglioside has no terminal sialic acid the Sendai virions do not attach [161]. It is presently unknown whether sialoglycoproteins or sialoglycolipids (gangliosides) are the actual cellular reception sites for myxoviruses, or if both function as such. It is suggested that it is sialoglycoproteins which are more likely to contact myxovirions first. Haywood's proposal (Sendai virus first adsorbs to the outer glycoprotein, then the viral neuraminidase causes the virus's release so that it can in turn attach to membrane gangliosides before penetration) does not explain why other myxoviruses treated with antineuraminidase (or those viruses without neuraminidase, e.g. measles) retain infectivity [151d, 153]. However, Woodruff and Woodruff [162] have obtained results suggesting that while sialic acid is the common element of the myxovirion attachment sites on lymphocytes, there are separate reception sites for ortho- and paramyxovirions, and that the reception sites for NDV and Sendai virus are also distinct. Thus the sialic acid determinants on susceptible host cells are probably different for different myxoviruses, and may, for example, incorporate oligosaccharide sequences containing other sugars in addition to sialic acid. Simple extrapolations are not therefore valid.

4.4.4 Other large enveloped viruses

Rather less is known regarding virus attachment for other groups of large enveloped viruses such as oncornaviruses (RNA tumour viruses), herpes viruses and poxviruses. The latter two classes have a DNA genome which can code for many (50–200) viral proteins, and the virions are structurally complex. For example, herpes simplex virus contains at least 24 structural polypeptides, 12 of which are probably glycosylated [163], a complexity which approaches that of many host cell surface membranes and makes the task of identifying the components involved in attachment yet more difficult. Vaccinia virus, a pox virus, has a genome of 160×10^6 daltons which is large enough to code for approximately 200 proteins with an average molecular weight of 40 000 [164].

Despite the structural complexity of these viruses they attach to susceptible host cells as fast as many smaller viruses [165, 166, 167], and it is probable that glycoproteins in the viral envelopes are involved in attachment.

Proof that viral glycoproteins are indeed required for attachment of oncornaviruses, which are not as structurally complex as herpes and poxviruses, has recently been obtained by De Largo and Todaro [168]. The 71 000 dalton envelope glycoprotein of Rauscher murine leukaemia virus was purified and shown to attach specifically to murine, but not other mammalian, cells in culture. Some 5×10^5 reception sites for the glycoprotein were estimated to exist on each mouse NIH/3T3 cell. Interestingly, the reception sites for this glycoprotein were blocked or absent in mouse cells producing related ecotropic type C viruses, and De Largo and Todaro suggest that the reception sites may be used for budding of progeny virus particles, and thus become depleted in these cells.

The biochemical nature of the reception sites for oncornaviruses is unknown. However, the control of the host range of avian tumour viruses has been studied extensively and appears to depend upon the existence of specific reception sites for each viral subgroup. A particular blood group antigen (R1) of chick erythrocyte, which may be related [252] to the human blood group M and N antigens (well known as important determinants of myxovirus adsorption), is associated with susceptibility to subgroup B virus, RAV-2 [226]. Different receptors on chick embryo fibroblasts were detected for subgroups (RAV–1) [227]. Like the murine leukaemia virus removal of the glycoprotein spikes enzymatically [228] or genetically [229] renders the virus non-infectious, presumably by preventing adsorption.

4.4.5 Summary

In summary, we can see that enveloped viruses have several features in the attachment process which are common to the group as a whole. All of the enveloped viruses have glycoproteins on their surface which form projections and which are in many cases necessary for attachment. Further, there is growing evidence at least in VSV that the viral glycoprotein oligosaccharides, or some unit or sequence of these, is critically involved in the recognition by cellular reception sites. On the other hand, in the cases of myxoviruses and the flavivirus, Japanese B encephalitis, it seems clear that carbohydrate residues (sialic acid, mannose, respectively) are present as integral components of the host cell reception sites. Thus, it is apparent that carbohydrate linked to proteinor lipid components both of viruses and host cells has an important role in determining the specifcity of the interaction between enveloped virus and cell surfaces (Fig. 4.1).

4.5 PHENOMENA ASSOCIATED WITH VIRUS ATTACHMENT

4.5.1 Haemagglutination

This phenomenon has been found in virtually all groups of enveloped viruses and many non-enveloped viruses when these viruses are mixed at sufficiently high concentration with certain types of mammalian or avian erythrocytes (red blood cells). In the process of haemagglutination large numbers of virions must first attach to the erythrocyte reception sites before the virus-erythrocyte lattice can form and the cells clump and settle out of solution. Divalency or multivalency of the viral recognition units or ‘ligands’ is an obvious requirement for haemagglutination. For example, intact adenoviruses can cause haemagglutination since they possess 12 fibre projections, and thus can form a 'bridge' between two erythrocytes. However, isolated fibre protein cannot haemagglutinate because it can only attach monovalently [5]. In some cases haemagglutination will occur with isolated viral glycoproteins, e.g. influenza haemagglutinin glycoprotein, but this is because they aggregate in solution and form structures (rosettes) that can bind multivalently [151].

The process of virus-induced haemagglutination may be compared with lectin agglutination of cells in suspension. Lectins are plant proteins and glycoproteins which bind specifically to certain carbohydrate units or sequences, e.g. concanavalin A binds to mannose [16]. Therefore, they will bind to cell surfaces since carbohydrate-containing molecules, glycoproteins, glycosaminoglycans and glycolipids, are abundant there. Since, in general, lectins are polyvalent proteins they can also bind cells together and cause agglutination. However, it is known that lectins, after attaching to their cell surface reception sites, induce a redistribution of reception site molecules in the surface membrane such that these sites cluster before agglutination occurs; this happens more readily in many transformed cells than in their normal cell counterparts [17]. Viral protein and glycoprotein recognition units can be likened to lectins in that they often have lectin-type specificity, e.g. influenza HA protein which binds to sialic acid, and cause agglutination of cells, especially erythrocytes, when present in intact virions or aggregates; i.e. in states where polyvalent attachment is possible. There is, however, relatively little free diffusion of surface proteins in erythrocyte plasma membranes because of a rigid cytoskeleton which acts transmembranally to restrict movement of antigens and prevent cell deformation [13]. Thus the virus–erythrocyte system more closely resembles a bacteriophage-bacterial cell interaction where phage recognition units attach to rigidly held cell-wall reception sites than the virus-susceptible host cell systems where viral recognition units attach to metastable, transient surface membrane sites in a 'fluid' membrane.

Compared to the surface membrane of other tissues and cultured cells, the erythrocyte plasma membrane is relatively simple. There are approximately 20 proteins and glycoproteins, but only two glycoproteins are present in large amounts on the cell surface [13]. These major surface glycoproteins are usually referred to as glycophorin [159] and Component III [13, 170]. Glycophorin is a sialoglycoprotein of apparent molecular weight 30 000–50 000, and is approximately 60% carbohydrate by weight [159, 169, 171, 172]. The oligosaccharides of glycophorin form the blood group antigens (ABO and MN), bind several different lectins, and are involved in virus attachment, e.g. influenza, NDV [159, 173–176]. Component III, a glycoprotein of molecular weight 90 000–100 000 has approximately 6–8% carbohydrate by weight and this specifically binds the lectin concanavalin A [170]. The erythrocyte plasma membrane is thus fairly atypical, and probably the reception sites for viruses or those molecules bearing reception sites, e.g. sialic acid for myxoviruses, are different to those found in susceptible host cells. Other evidence, for example, after chemical/biochemical modification the loss of ability of several viruses (enteroviruses, VSV, influenza) to haemagglutinate without a corresponding loss of infectivity for susceptible cells also suggests that virus–erythrocyte interactions can be different from virus-host cell attachment.

Biochemically, the process of haemagglutination for the majority of viruses has been little studied. Myxoviruses form a major exception. The erythrocyte reception sites for myxoviruses contain sialic acid and the major sialoglycoprotein glycophorin carries the bulk of these reception sites [176–178]. Haemagglutination with myxoviruses is optimal at 4 °C, since most of these viruses contain a neuraminidase which at 37 °C cleaves the terminal sialic acid residues to which virions are bound. The virions are then released or eluted from the erythocyte. Bacterial neuraminidases, which are available commercially, do not always destroy erythrocyte reception sites for myxoviruses [248]. Unlike the viral enzyme which prefers the $\alpha 2 \rightarrow 3$ linked sialic acid sequence, bacterial neuraminidases, e.g. *Corynebacterium diphtheriae,* hydrolyse other linkages more efficiently, implying that the natural receptor site for myxoviruses involves predominantly the $\alpha 2 \rightarrow 3$ linkage. The sialic acid reception sites for myxoviruses are also inactivated by periodate treatment followed by borohydride reduction which removes carbon atoms 7, or 8 and 9 from the sialic acid residues [179], and therefore indicates the requirement for intact sialic acid molecules in myxovirus attachment. Several purified sialoglycoproteins, e.g. fetuin, and sialoglycolipids (gangliosides) [158, 178, 217] as well as solubilized glycophorin, inhibit haemagglutination by myxoviruses and detailed evidence for the specificity of binding has come from measurements of the ability of these substances to inhibit viral haemagglutination (Table 4.2). In the first place, a high concentration of terminal residues of sialic acid are absolutely necessary for the interaction. Thus, colominic acid, a polymer of *N*-acetylneuraminic acid was inactive in preventing haemagglutination [178, 217] presumably because of the very low content of non-reducing

Table 4.2 Inhibition of influenza virus haemagglutination by soluble glycoproteins

Glycoprotein	Mol. wt. x 10^4	Sialic acid moles/mole	Peripheral sequence	Inhibitory activity (%)†
Human MM substance	1200	4466	NeuNAc NeuNAC* (α2, 6) (α2,6) gal (β13)–gal NAc-	100.0
Bovine submaxillary mucin	225	1667	NeuNAc (α2, 6) galNAc-	7.5
Ovine submaxillary mucin	75	680	NeuNAc (α2, 6) galNAc-	3.0
Human NN substance	59	312	As MM substance*	4.2
Human ceruloplasmin	16	12	NeuNAc (α2, 6) gal-	< 0.12
Bovine fetuin	4.5	9	NeuNAc (α2, 6) gal (β1, 4) glcNAc-	0.24
Human orosomucoid	4.4	17	As fetuin	< 0.12
Human MM substance subunit	3.4	13	As MM substance*	1.0

* See Springer and Desai [252] for more recent assignments.

† Viral haemagglutination inhibition data adapted from Springer [217]. Activity relative to human MM substance on an equal weight basis.

sialic acid residues. Secondly, an intact sialic acid non-reducing residue is necessary since selective oxidation of the acyclic side chain at C7–C9 of sialic acid terminals of glycoproteins abolished their inhibitory activity [179, 218]. Simultaneously, the susceptibility of these modified terminal residues to viral neuraminidase is also markedly decreased. These results show, therefore, that the polyhydroxy 3-carbon fragment of sialic acid plays an important role in binding by the viral spike glycoprotein. Presumably, the carboxyl function of sialic acid also participates in the binding. Some electrostatic interaction between viral haemagglutinin and sialic acid residues is indicated by the reversibility of binding, carried out in solutions of moderate electrolyte concentration simply by raising the salt concentration.

In addition to an intact sialic acid non-reducing terminal residue, some strains of influenza virus show greater specificity requirements. Thus, the binding of the Asian (A2) strain of influenza virus is strongest to sialic acid residues bearing an *O*-acetyl substituent at C4 [218]. On the other hand, there appears to be relatively little specificity for the sugar penultimate to the terminal sialic acid residue. This is shown, for example, by the similar inhibitory activities of ovine submaxillary mucin and serum glycoproteins in which sialic acid residues are attached respectively to *N*-acetylgalactosamine and galactose. However, as well as a simple α-ketosidically linked sialic acid residue other components are involved in virus binding. Simple glycosides such as the α-methyl glycoside or the disaccharide prosthetic group of ovine submaxillary gland mucin, sialyl (α2, 6) *N*-acetylgalactosamine are very poor inhibitors of viral haemagglutination. The virus-inhibiting potential of soluble glycoproteins, such as ovine submaxillary gland mucin, therefore depends critically on the way in which the sialic acid-containing side chains is presented to the virus, particularly on the molecular size and configuration of the intact glycoprotein [217]. Low molecular weight sialoglycopeptides produced from ovine submaxillary gland mucin by trypsin digestion are devoid of inhibitory activity [219]. Conversely, a relatively poor inhibitor such as human α_1-acid glycoprotein can be converted into a potent inhibitor of viral haemagglutination by artifically increasing its molecular size. This can be done by heating at pH 3.8 and 80 °C to form on average pentameric aggregates [220]. Alternatively, the glycoprotein is cross-linked with acetaldehyde to produce large aggregates [221]. The treated glycoproteins are at least 500 times more powerful in inhibiting haemagglutination using PR8 influenza as the indicator virus.

Ortho- and paramyxoviruses appear to have similar erythrocyte sialic acid reception sites and both types of viruses elute from erythrocytes at 37°C because of terminal sialic acid cleavage by virion neuraminidase, paramyxoviruses more slowly. However, other viruses which have been included with myxoviruses such as measles, canine distemper and rhinderpest (now called pseudomyxoviruses) and which do not possess virion neuraminidase probably have different erythrocyte reception sites. Formaldehyde-treated erythrocytes were shown to have reception sites for ortho- and paramyxoviruses, but reception sites for measles virus were inactivated [180]. Also, pseudomyxoviruses appear to haemagglutinate only primate erythrocytes, in contrast to ortho- and paramyxoviruses which haemagglutinate fowl and various other mammalian (including primate) erythrocytes [181]. The erythrocyte reception sites for avian corona viruses [185, 227], polyoma virus [72, 76], cardioviruses (encelphalomyocarditis virus) [186, 187] and an equine rhinovirus [41] are all apparently inactivated by neuraminidase. However, not all erythrocyte reception sites are affected by neuraminidase, particularly those for human rhinoviruses, poliovirus [41], rabies virus [188], adenovirus type 7 [102, 103] and the three serotypes of reovirus [71]. However, in all of the above examples no distinction can be made as to whether attachment is blocked, or whether some undefined property of the virions necessary for the formation of a virus-erythrocyte lattice is inactivated.

In common with myxovirus attachment to susceptible host cells, it is the viral envelope glycoproteins which are involved in haemagglutination. Whilst there is no evidence that the carbohydrate of these molecules acts in the interaction with host cells (contrast with VSV), there is a suggestion that this is possibly so in haemagglutination. For instance, Bikel and Knight [182] have shown that the PR8 influenza strain haemagglutinin, but not that of NDV, was inactivated by treatment with a purified, protease-free glycosidase mixture containing *N*-acetylglucosaminidase. Also, as mentioned previously, sialylated influenza virions lose the ability to haemagglutinate [151d], although infectivity is not affected.

Stronger evidence for a role for virion carbohydrate in erythrocyte attachment is available for other enveloped viruses, particularly VSV and some alphaviruses. VSV has terminal sialic acid residues on its glycoprotein oligosaccharides which are the viral recognition units for attachment to host cells [119, 121, 125]. Neuraminidase treatment of VS virions also leads to a loss of ability to haemagglutinate goose erythrocytes [125], which is restored on re-sialylation. In contrast, removal of sialic acid from Semliki Forest or Sindbis virions does not inactivate haemagglutination by these viruses [136, 138]. However, further sequential degradation of alphavirus glycoprotein carbohydrate chains or growth of virus in cells lacking *N*-acetylglucosaminyl transferase activity, where the viral glycoproteins are under-glycosylated (Fig. 4.3), yields virions which do not haemagglutinate [127, 138]. Since the latter mutant cells yield progeny virus which is fully infectious [127, Meager, unpublished results], it is suggested that attachment may be unaffected, and that those virions lacking complete carbohydrate chains are unable to cause haemagglutination for other reasons. When most of the carbohydrate is removed [138] the virions aggregate badly, and this probably affects their ability both to infect host cells and to cause haemagglutination.

There is also some evidence that carbohydrate attached to non-enveloped viruses is involved in haemagglutination. Periodate or sodium borohydride, for example, have been shown to inhibit haemagglutination by enteroviruses and reoviruses [71, 182, 183]. However, the cited reagents affect other groups besides *cis*-glycols (KIO_4) and carboxyl groups ($NaBH_4$), e.g. serine (KIO_4) and disulphide bonds ($NaBH_4$), and hence the involvement of virion carbohydrate in such cases is questionable [182, 184]. This contention is supported, at least for enteroviruses, by the lack of unequivocal results proving carbohydrate to be an integral part of enterovirions [41]. A very recent report [67] demonstrating the presence of carbohydrate in reovirus capsid proteins does nevertheless support evidence obtained in earlier investigations that reovirion carbohydrate was necessary for haemagglutination [71, 182, 183]. Generally speaking, however, the proposal that carbohydrate, believed to be covalently attached to non-enveloped viruses, has a crucial role in both virus attachment and haemagglutination is, in the authors opinion, tenuous.

In the case of adenoviruses, the fibre protein projection may contain carbohydrate [41, 95], but Neurath *et al.* [102, 103] have found that 2, 3-butanedione which reacts with arginine inhibits haemagglutination by type 7 adenovirus and suggest that arginine present in the knob of the fibre projection forms a salt bond to a negative charge on the erythrocyte surface. There are, however, many different adenoviruses, and it is as yet unknown whether fibre arginine residues for all serotypes are necessary for attachment to erythrocytes. In the case of adenovirus type7 the reception sites on human erythrocytes can be destroyed by 1-cyclohexyl-3-(2-morpholinoethyl)-carbodimidemetho-*p*-toluenesulphonate, a reagent which modifies carboxyl groups [103]. This suggests that protein carboxyl groups are present in the reception sites, and it has been hypothesised that these form a salt bond with arginine residues present on the adenovirus fibre projections.

To summarize, therefore, while haemagglutination has proven valuable in defining the specificity of myxovirus binding to cells, the relationship between agglutination of erythocytes and interactions with host cells supporting viral replication is more questionable for other viruses, particularly in the class of non-enveloped viruses. Whether such a relationship exists will depend on more detailed structure analysis of the reception site molecules of host cells for comparison with the erythrocyte membrane components.

4.5.2 Virus entry

Once virions are attached to the surface of cells at their reception sites, the next step is passage of their genetic information into the cell cytoplasm. To do this the virion must 'penetrate' the cell plasma membrane. Our observations of virus-cell systems show us that virus penetration occurs by two principal methods. These are (a) viropexis and (b) fusion.

Viropexis

Viropexis is a process similar to phagocytosis or endocytosis, and means that the attached virion(s) (assuming that they are attached to the correct reception sites on the host cell) are engulfed by the plasma membrane, and are internalized in vesicles (Fig. 4.5). Why this occurs is unknown, but some suggestions are made below. It seems that even simple membranes, such as ganglioside liposomes 'attempt' to engulf attached virions, e.g. Sendai virus [188], at 37°C. Viropexis (endocytosis) is probably an inherent function of most host cell plasma membranes, e.g. silica beads are rapidly taken up by cells in culture, and thus many viruses are taking advantage of or exploiting this natural occurrence. The process is temperature-dependent and at low temperatures, 0–4°C, virions can attach to the cell surface but are not internalized. Viropexis is rapid at 37°C. Once virions have entered the host cell in vesicles or vacuoles they appear to be able to penetrate out of these, perhaps after contact with lysosomes (although this point is hotly disputed [153]), with further breakdown of the virion capsid. Ultimately the process of virus replication begins when the viral nucleic acid, or the cores containing this, are free in the cytoplasm and can subvert biosynthetic machinery of the host cell.

Under the electron microscope, virtually all groups of viruses, both enveloped and non-enveloped, have been observed entering susceptible

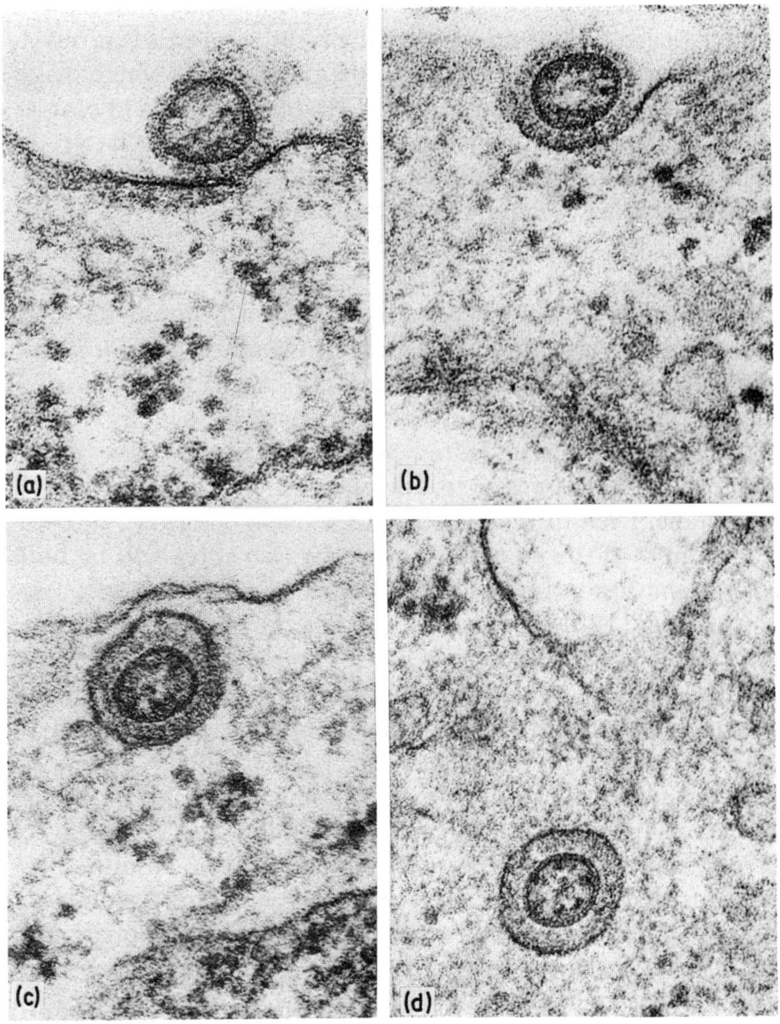

Fig. 4.5 (a–d) Electron micrographs showing stages in attachment and penetration of myxovirus particles (fowl plague virus) into chick embryo fibroblasts, seen 15 minutes after virus inoculation x 165 000 approx. Kindly taken and supplied by Dr J.A. Armstrong.

host cells by viropexis. Dales [83] lists the following viruses entering cells by viropexis: herpes viruses, poxoviruses, adenoviruses, papovaviruses, reoviruses, picornaviruses, rhabdoviruses, myxoviruses and oncornaviruses (RNA tumour viruses). For non-enveloped viruses, viropexis is probably the sole method of entry, although there are a few reports where the authors claim that the virions (or degraded virions [242–244]) pass directly through the plasma membrane [189, 230, 237]. The latter process is difficult to envisage, and hard to justify from the few electron micrographs purporting to demonstrate this.

Fusion

For enveloped viruses there is a second method by which the virions, or their cores, gain entry, and this is by fusion of the virion glycoprotein-phospholipid envelope with that of the host cell plasma membrane [232]. Nevertheless, of the enveloped viruses only some of the paramyxovirus group, notably Sendai virus, and a few members of the arbovirus (togavirus) group [231], are able to fuse efficiently with the host cell plasma membrane. Even these enter by viropexis on occasion [190–192]. Vaccinia, a member of the poxvirus group, may also enter cells by both viropexis [83] and fusion [167, 193], although the more recent work by Chang and Metz [193] favours fusion as the principal method of vaccinia entry (Fig. 4.6).

The basic mechanism of fusion remains unexplained and both membrane lipids [233] and membrane proteins [234] have been proposed as the major factors in fusion. Recently, Lucy has combined these mechanisms in a new theory of fusion [235] which may also be applicable in enveloped virus-host cell interactions. Studies with synthetic protein-free membranes (liposomes) seem to eliminate an absolute requirement for host cell membrane proteins in membrane fusion by Sendai virus [236]. The viral spike glycoproteins, however, appear to play an important role since the liposomes contained ganglioside receptor sites. Subsequent events may, however, be determined by the non-ganglioside lipid composition of the liposomes. When Sendai virions were adsorbed to liposomes containing only phosphatidyl choline, cholesterol and gangliosides and incubated at 37°C for several hours, the liposomes and viral membranes did not fuse. The liposomal membrane appeared to flow around each virion rather as in a phagocytic engulfment. However, addition of phosphatidylethanolamine and sphingomyelin to phosphatidylcholine, cholesterol and gangliosides resulted in liposomes that fused with the virus. The requirement for specific lipids is extremely interesting and may indicate recognition by the viral spike glycoprotein. Alternatively, specific lipids may be required to allow a suitable

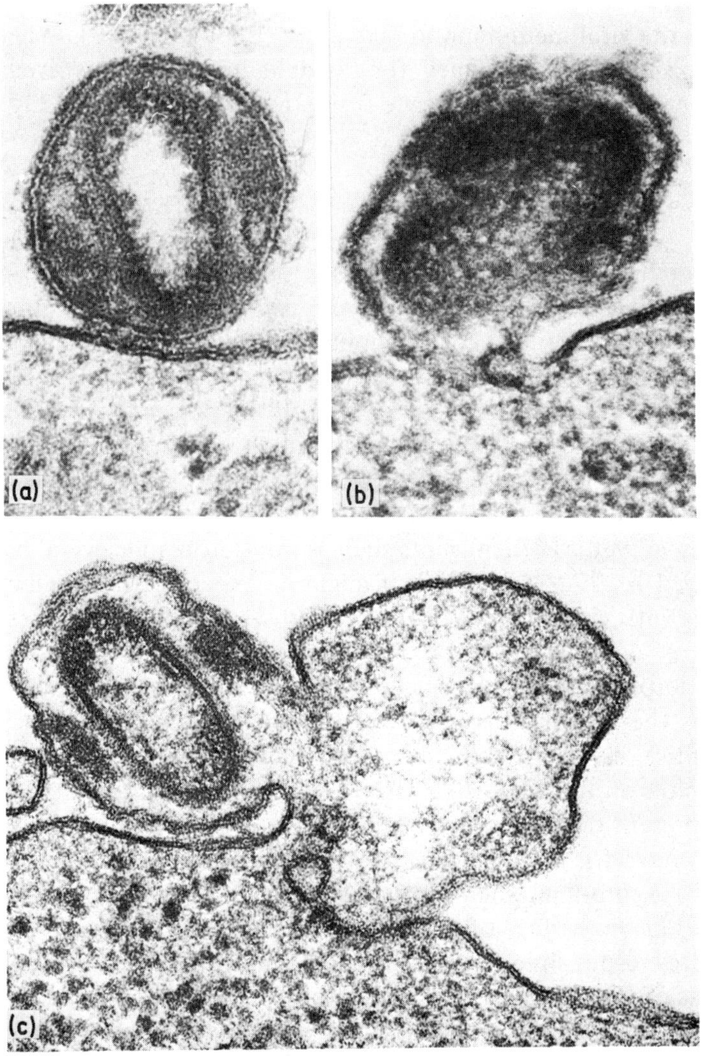

Fig. 4.6 Entry of a poxvirus (vaccinia) into cultured L cells by means of membrane fusion 15 minutes after addition of the virus x 140 000 approx. (a) Vaccinia absorbed to the cell surface. (b) Commencing fusion of viral and cell membranes. (c) Continuity of viral envelope and cell membrane after fusion. Electron micrographs [167] kindly supplied by Dr J.A. Armstrong.

perturbation of the lipid bilayer either in a protein-free liposome, or a protein-free patch on the host cell surface in contact with virus, for fusion with the viral membrane to take place [235, 239, 240]. Whether or not endogenous viral enzymes (for example, lipases) are involved in addition in virus-host cell membrane fusion remains debatable [237].

4.6 SUMMARY AND GENERAL COMMENTS

The first step in viral replication involves interaction of the virus particles with the surface of the host cell. It is presumed that specific receptor sites are present on the host cell membranes although only in one or two cases has this been shown unequivocally. In the best studied case of myxovirus adsorption the sialic acid receptors are present in glycoproteins and glycolipids of the host cell plasma membrane. Since carbohydrate groups are commonly exposed at the external surface of mammalian cells and are important reception sites for hormones such as insulin or thyrotropin, as well as other agents such as interferon and several bacterial toxins [139, 140, 222–224, 241], it might be expected that similar interactions with viruses other than myxoviruses would involve these substances.

For the purposes of the present discussion, the animal viruses may be classified by the presence or absence of a viral envelope. In both classes, viral adsorption involves complementary structures, namely the reception (receptor) sites on the host cell surface and the viral component (recognition unit) present on the surface of the virion. In the eveloped viruses, the recognition units for the reception sites must themselves be part of components of a membrane. The haemagglutinin glycoprotein of myxoviruses forms spikes at the surface of the virion as discussed previously and these carry the sites responsible for binding to sialic acid groups of membrane components of the host cell. In a non-enveloped virus such as the adenoviruses, the attachment to the host cell involves a structural protein projecting from the surface of the virion analogous to the tail fibres of T4 phages and the mode of attachment to the host cell is similar. It is interesting that in all viruses except the simplest, attachment seems to occur through projections from the viral surface. Pilation is also an important factor in the attachment of other micro-organisms to host cells [238]. Perhaps the initial contact to the host cell surface is made by projections because of the smaller surface area of such structures which overcomes the electrostatic barrier of similarly charged interacting surfaces. Preliminary attachment by a projection would ensure that the two surfaces had a maximum separation equal to the length of the projection and so would increase the probability of a closer approach. A close approach would favour attachment as the attractive forces produced by hydrophobic interactions or London dispersion forces act over a shorter range than electrostatic repulsive forces, leading to a stable adhesive bond at a critical separation where the potential energy is at a secondary minimum [200].

Is any great amount of specificity required for virus attachment? In the majority of cases the answer is apparently, no. The enveloped viruses for example, attach to virtually any cell they collide with, and this has the advantage for the virus in that all cellular organisms are (unwilling) prospective hosts. Viruses will also attach to non-biological surfaces, e.g. glass. However, the rapidity with which viruses attach to cells does suggest that there is a higher probability of successful collision in the latter case. In other words, there is a good finite probability of finding configurational complementarity between the virion surface recognition units and reception sites at the cell surface.

Other results lead to the conclusion that productive binding of a virus such as influenza virus or Sendai virus to host cell surfaces does not simply involve complementary groupings on the virus envelope glycoprotein and the host cell reception site(s). In addition, the number of such contacts made between each virion and the combining surface are important. The tightness of binding of myxovirus relies to a large extent on the number of sialic acid residues available per glycoprotein molecule or per unit area of surface. These contacts on the host cell surface would presumably be separated by approximately 7 nm which is the average distance between the spikes on the viral surface. So far as the soluble inhibitory glycoproteins of myxovirus-induced haemagglutination (Table 4.2) are concerned, the most important aspect of their structure seems to be to provide a template similar to that postulated as existing at adsorbing cell surfaces. For example, in ovine submaxillary gland mucins, prosthetic groups containing sialic acid may be separated by a distance about equal to the centre-to-centre distance between spike glycoproteins of the viral envelope.

Further information on the likely area of contact made by viruses adsorbed to cell membranes is provided by studies using synthetic lipid membranes. Haywood shows clearly that there is a minimum surface density of ganglioside reception sites required for efficient attachment of Sendai virions to synthetic membranes. Free gangliosides existing as

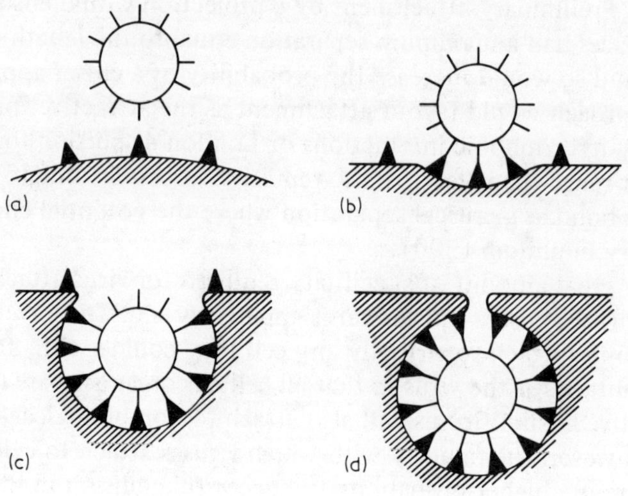

Fig. 4.7 (a–d). Hypothetical mechanisms of attachment of a complex (multivalent) virion with specific surface receptors and subsequent engulfment.

simple micelles in aqueous solution do not inhibit viral haemagglutination, whereas mixed ganglioside-phospholipid liposomes of greater surface area are potent inhibitors. The surface area available to attaching virions is critical since sonication to form smaller mixed vesicles containing a similar proportion of gangliosides reduces their capacity to inhibit viral haemagglutination dramatically. Therefore, the gangliosides are held in liposomes (as in natural membranes) at relatively high density over quite extensive areas of the available surface and are able to orientate themselves locally in order to make multiple points of contact with each virion. As Haywood points out, the local concentration of gangliosides may be increased by a capping or clustering mechanism (a temperature-dependent step) induced by the multivalent nature of the virion surface, carrying as it does many spikes. Although each haemagglutinin molecule is monovalent as shown by the inability of the glycoprotein isolated from influenza virus after detergent extraction or release by trypsin to agglutinate erythrocytes, each virion contains many such molecules and is multivalent. Under these conditions, therefore, each virion attached to the membrane surface may sequester many ganglioside molecules, and keep them in a fixed topography at the surface. Indeed, such a mechanism (Fig. 4.7) may be necessary to amplify the weak binding of each viral haemagglutinin and may even facilitate engulfment. If clustering of ganglioside viral reception sites occurs, then some local destabilization of the membrane would be expected in these areas enriched in gangliosides [242]. It is possible in this case that the local concentration of gangliosides would facilitate fusion of the viral envelope and the host cell membrane and aid in viral penetration, another possible next step in the replicative cycle. It would be intriguing to know to what extent this process also aids the fusion of cells and formation of cell hybrids through mediation of an enveloped virus, such as Sendai virus, in view of the probable importance of mobile lipid units in opposing membranes of fusing cells.

The electron micrograph pictures of the specific attachment of Sendai virus strongly indicate that each virion makes contacts with the membrane through a large number of surface spikes. Only rarely is a virus seen making contacts with one or two spikes. In general, the length of contact appears to be about 210–250 nm. If the area of attachment is assumed to be a circle of this diameter then the area involved could include as many as 3000 ganglioside molecules. It is not known, of course, how many contacts are actually made but presumably an upper limit of a few hundred is indicated by the estimated number of haemagglutinin molecules per virion. The calculation, however, is in general agreement with the conclusions made earlier concerning multiple points of contact between the virus and the host cell surface and the importance of this mechanism in subsequent steps.

REFERENCES

1. Crick, F.H.C. and Watson, J.D. (1956), *Nature*, **177**, 473–475.
2. Nermut, M.V. (1975), *Virology*, **65**, 480–495.
3. Schultze, I.T. (1973), *Adv. Virus Res.*, **18**, 1–55.
4. Hughes, R.C. (1973), *Prog. Biophys. Mol. Biol.*, **26**, 189–268.
5. Norrby, E. (1968), *J. gen. Virol.*, **5**, 221–236.
6. Blough, H.A. and Tiffany, J.M. (1975), *Current Topics in Microbiology and Immunology*, **70**, 1–30.
7. Wallach, D.F.H. (1972), *The Plasma Membrane: Dynamic Perspectives Genetics and Pathology*, English Universities Press, London.
8. Jamieson, G.A. and Robinson, D.M. (eds.), (1976), *Mammalian Cell Membranes*, Vol. 1. General Concepts, Butterworths, London-Boston.
9. Chapman, D. and Wallach, D.F.H. (eds.), (1973), *Biological Membranes*, Vol. 2, Academic Press, London-New York.
10. Hughes, R.C. (1976), *Membrane Glycoproteins*, Butterworths, London-Boston.
11. Singer, S.J. and Nicolson, G.L. (1972), In: *Membranes and Viruses in Immunopathology*, Day, L, and Good, R.A. (eds.), Academic Press, New York, pp. 7–47.

12. Singer, S.J. and Nicolson, G.L. (1972), *Science,* **175**, 720–731.
13. Nicolson, G.L. (1976), *Biochim. biophys. Acta,* **457**, 57–108.
14. Jamieson, G.A. and Greenewalt, T.J. (eds.), (1970), *The Red Cell Membrane,* Lippincott, Philadelphia.
15. Evans, W.H. (1976), *Biochem. Soc. Transactions,* **4**, 1009–1013.
16. Lis, H. and Sharon, N. (1973), *Ann. Rev. Biochem.,* **42**, 541–559.
17. Nicolson, G.L. (1976), *Biochim. biophys. Acta,* **458**, 1–72.
18. Rapin, A.M.C. and Burger, M.M. (1974), *Adv. Cancer Res.,* **20**, 1–91.
19. Koch, A.S. and Feher, G.Y. (1974), *Acta Microbiol. Acad. Sci., Hung.,* **21**, 245–256.
20. Bachrach, H.L. (1968), *Ann. Rev. Microbiol.* **22**, 201–244.
21. Philipson, L., Beatrice, S.T. and Crowell, R.L. (1973), *Virology,* **54**, 69–79.
22. Rueckert, R.C. (1971), In: *Comparative Virology,* (Maramorosch, and Kurstak, (eds.), Academic Press, New York, pp. 255–300.
23. Crowell, R.L. and Philipson, L. (1971), *J. Virol.,* **8**, 509–515.
24. Lonberg-Holm, K. and Korant, B.D. (1972), *J. Virol.,* **9**, 29–40.
25. Roesing, T.G., Toselli, P.A. and Crowell, R.L. (1975), *J. Virol.,* **15**, 654–667.
26. Breindl, M. and Koch, G. (1972), *Virology,* **48**, 136–144.
27. Crowell, R.L., Landau, B.J. and Philipson, L. (1971), *Proc. Soc. exp. Biol. Med.,* **137**, 1082–1088.
28. Fenwick, M.L. and Cooper, P.D. (1962), *Virology,* **18**, 212–223.
29. Lonberg-Holm, K. and Philipson, L. (1974), Early interaction between animal viruses and cells. In: *Monographs in Virology,* Vol. 9, (Melnick, J.L., ed.), S. Karger, Basel, pp. 4–21.
30. Holland, J.J. (1964), *Bact. Rev.,* **28**, 3–13.
31. McClaren, L.C., Holland, J.J. and Syverton, J.T. (1959), *J. exp. Med.,* **109**, 475–485.
32. Kumin, C.M. (1962), *J. Immunol.,* **88**, 556–569.
33. Kumin, C.M. (1964), *Bact. Rev.,* **28**, 382–390.
34. Kumin, C.M. and Halmagyl, N.E. (1961), *J. Clin. Invest.,* **40**, 1055–1056.
35. Baron, S., Friedman, R.M. and Buckler, C.E. (1963), *Proc. Soc. exp. Biol. Med.,* **113**, 107–110.
36. Holland, J.J. (1961), *Virology,* **15**, 312–326.
37. Soloviev, V.D., Krispin, T.I., Zaslavsky, V.G. and Agol, V.I. (1968), *J. Virol.,* **2**, 553–557.
38. Miller, D.A., Miller, O.J., Dev, V.G., Hashmi, S., Tantravashi, R., Medrano, L. and Green, H. (1974), *Cell,* **1**, 167–173.
39. Crowell, R.L. (1966), *J. Bact.,* **91**, 198–204.
40. Medrano, L. and Green, H. (1973), *Virology,* **54**, 551–524.
41. Lonberg-Holm, K. and Philipson, L. (1974), Early interactions between animal viruses and cells. In: *Monographs in Virology,* (Melnick, J.L, ed.), Vol. 9, S. Karger, Basel, pp. 24–49.

42. Quersin-Thiry, L. and Nihoul, E. (1961), *Acta Virol., Praha,* **5**, 283–293.
43. Lonberg-Holm, K., Crowell, R.L. and Philipson, L. (1976), *Nature,* **259**, 679–681.
44. Levitt, N.H. and Crowell, R.L. (1967), *J. Virol.,* **1**, 693–700.
45. Zajac, I. and Crowell, R.L. (1965), *J. Bact.,* **89**, 1097–1100.
46. Holland, J.J. and McClaren, L.C. (1959), *J. exp. Med.,* **109**, 487–504.
47. Holland, J.J. and McClaren, L.C. (1961), *J. exp. Med.,* **114**, 167–171.
48. Quersin-Thiry, L. (1961), *Acta Virol., Praha,* **5**, 141–152.
49. Thorne, H.V. (1963), *J. Bact.,* **85**, 1247–1255.
50. Chan, J.C. and Black, F.L. (1970), *J. Virol.,* **5**, 309–312.
51. McClaren, L.C., Scaletti, J.V. and James, C.G. (1968), In: *Biological Properties of the Mammalian Surface Membrane,* (Manson, L., ed.), Wister Institute Press, Philadelphia, pp. 123–135.
52. Philipson, L. and Choppin, P.W. (1962), *Virology,* **16**, 405–413.
53. Allison, A.C., Buckland, F.E. and Andrewes, C.H. (1962), *Virology,* **17**, 171–175.
54. Buckland, F.E. (1960), *Nature,* **188**, 768.
55. Philipson, L. and Choppin, P.W. (1960), *J. exp. Med.,* **112**, 455–478.
56. Holland, J.J. (1962), *Virology,* **16**, 163–176.
57. Lonberg-Holm, K. and Butterworth, B.E. (1976), *Virology,* **71**, 171–175.
58. Carthew, P. and Martin, S.L. (1974), *J. gen. Virol.,* **24**, 525–534.
59. Talbot, P., Rowlands, P.J., Burroughs, J.N., Sangar, D.V. and Brown, F. (1973), *J. gen. Virol.,* **19**, 369–380.
60. Butterworth, B.E., Grunert, R.R., Korant, B.D., Lonberg-Holm, K. and Yin, F.H. (1976), *Arch. Virol.,* **51**, 169–189.
61. Much, D.H. and Zajac, I. (1974), *J. gen. Virol.,* **23**, 205–208.
62. Lonberg-Holm, K. (1975), *J. gen. Virol.,* **28**, 313–327.
63. Mandel, B. (1967), *Virology,* **31**, 238–247.
64. Heberling, R.L. and Cheever, F.S. (1965), *Proc. Soc. exp. Biol. Med.,* **120**, 127–130.
65. Noble-Harvey, J. and Lonberg-Holm, K. (1975), *J. gen. Virol.,* **25**, 83–91.
66. Smith, R.E., Zweerink, H.J. and Joklik, W.K. (1969), *Virology,* **39**, 791–810.
67. Krystal, G., Perrault, J. and Graham, A.F. (1976), *Virology,* **72**, 308–321.
68. Seehafer, J., Salmi, A., Scraba, D.G. and Colter, J.S. (1975), *Virology,* **66**, 192–205.
69. Wright, P.J. and di Mayorca, G. (1975), *J. Virol.,* **15**, 828–835.
70. Wright, P.J. and di Mayorca, G. (1976), *J. Virol.,* **19**, 750–755.
71. Lerner, A.M., Cherry, J.D. and Finland, M. (1963), *Virology,* **19**, 58–65.
72. Mori, R., Schieble, J.H. and Ackermann, W.W. (1962), *Proc. Soc. exp. Biol. Med.,* **109**, 685–690.
73. Halperen, S., Schieble, J.H., Mori, R., Dinka, S. and Ackermann, W.W. (1962), *J. Immunol.,* **89**, 400–407.
74. Mann, J.J., Rossen, R.D., Lehrich, J.R. and Kasel, J.A. (1967), *J. Immunol.,* **98**, 1136–1142.

75. Schmidt, N.J., Dennis, J., Hoffman, M.N. and Lennette, E.H. (1964), *J. Immunol.*, **93**, 367–376.
76. Hartley, J.W., Rowe, W.P., Chanock, R.M. and Andrews, B.E. (1959), *J. exp. Med.*, **110**, 81–91.
77. Gomatos, P.J. and Tamm, I. (1962), *Virology*, **17**, 455–461.
78. Dales, S., Gomatos, P.J. and Hsu, K.C. (1965), *Virology*, **25**, 193–211.
79. Loh, P.C., Soergel, M. and Oie, H. (1967), *Arch. Gesante Virusforsch.*, **22**, 398–408.
80. Silverstein, S.C. and Dales, S. (1968), *J. Cell Biol.*, **36**, 197–230.
81. Barbant-Brodano, G., Swetly, P. and Koprowski, H. (1970), *J. Virol.*, **6**, 78–86.
82. Hummeler, K., Tomassini, N. and Sokol, F. (1970), *J. Virol.*, **6**, 87–93.
83. Dales, S. (1973), *Bact. Rev.*, **37**, 103–135.
84. Basilico, C. and di Mayorca, G. (1974), *J. Virol.*, **13**, 931–934.
85. Andrewes, C. and Pereira, H.G. (1972), In: *Viruses of Vertebrates*, Bailliere-Tindall, London.
86. Valentine, R.C. and Pereira, H.G. (1965), *J. Mol. Biol.*, **13**, 13–20.
87. Norrby, E. (1966), *Virology*, **28**, 236–243.
88. Pettersson, U., Philipson, L. and Hoglund, S. (1968), *Virology*, **35**, 204–215.
89. Norrby, E. (1969), *J. Gen. Virol.*, **5**, 221–236.
90. Philipson, L., Lonberg-Holm, K. and Pettersson, U. (1968), *J. Virol.*, **2**, 1064–1075.
91. Hughes, R.C. and Mautner, V. (1973), In: *Membrane-Mediated Information* (Kent, P., ed.), Vol. 1, Medical and Technical Publishing Co., Lancaster, pp. 104–125.
92. Brown, D.T. and Burlingham, B.T. (1973), *J. Virol.*, **12**, 386–396.
93. Sundquist, B., Pettersson, U., Thelander, L. and Philipson, L. (1973), *Virology*, **51**, 252–256.
94. Dorsett, P.H. and Ginsburg, H.S. (1975), *J. Virol.*, **15**, 208–216.
95. Ishibashi, M. and Maizel, J.V. Jr. (1974), *Virology*, **58**, 345–361.
96. Meager, A., Butters, T.D., Mautner, V. and Hughes, R.C. (1976), *Eur. J. Biochem.* **61**, 345–353.
97. Meager, A., Nairn, R. and Hughes, R.C. (1975), *Virology*, **68**, 41–57.
98. Lonberg-Holm, K. and Philipson, L. (1969), *J. Virol.*, **4**, 323–328.
99. Boulanger, P.A., Houdret, N., Scharfman, A. and Lemay, P. (1972), *J. gen. Virol.*, **16**, 429–434.
100. Butters, T.D. and Hughes, R.C. (1974), *Biochem. J.*, **140**, 469–478.
101. Butters, T.D. and Hughes, R.C. (1975), *Biochem. J.*, **150**, 59–69.
102. Neurath, A.R., Hartzell, R.W. and Rubin, B. (1969), *Nature*, **221**, 1069–1071.
103. Neurath, A.R., Hartzell, R.W. and Rubin, B. (1970), *Virology*, **42**, 789–793.
104. Habel, K., Hornibrook, J.W., Gregg, N.C., Silverberg, R.J. and Takemoto, K.K. (1958), *Virology*, **5**, 7–29.
105. Pereira, H.G. (1958), *Virology*, **6**, 601–611.

106. Everett, S.F. and Ginsberg, H.S. (1958), *Virology*, **9**, 770–773.
107. Kjellen, L. (1972), *Arch. Gesamte Virusforsch,* **39**, 1–12.
108. Ankerst, J. and Kjellen, L. (1973), *Arch. Gesamte Virusforsch,* **41**, 99–105.
109. Russell, W.C., Valentine, R.C. and Pereira, H.G. (1967), *J. gen. Virol.,* **1**, 509–522.
110. Cartwright, B., Smale, C.J. and Brown, F. (1969), *J. gen. Virol.,* **5**, 1–10.
111. Wagner, R.R., Prevec, L., Brown, F., Summers, D.F., Sokol, F. and MacLeod, R. (1972), *J. Virol.,* **10**, 1228–1230.
112. Etchison, J.R. and Holland, J.J. (1974), *Virology,* **60**, 217–229.
113. McSharry, J.J., Compans, R.W. and Choppin, P.W. (1971), *J. Virol.,* **8**, 722–729.
114. McSharry, J.J. and Wagner, R.R. (1970), *J. Virol.,* **7**, 412–415.
115. Mudd, J.A. and Summers, D.F. (1970), *Virology,* **42**, 328–340.
116. Etchison, J.R. and Holland, J.J. (1974), *Proc. natn. Acad. Sci., U.S.A.,* **71**, 4011–4014.
117. Moyer, S.A. and Summers, D.F. (1974), *Cell,* **2**, 63–70.
118. Moyer, S.A., Tsang, J.M., Atkinson, P.H. and Summers, D.F. (1976), *J. Virol.,* **18**, 167–175.
119. Schloemer, R.H. and Wagner, R.R. (1975), *J. Virol.,* **15**, 1029–1032.
120. Sefton, B.W. (1976), *J. Virol.,* **17**, 85–93.
121. Schoemer, R.H. and Wagner, R.R. (1974), *J. Virol.,* **14**, 270–281.
122. Bussereau, F., Cartwright, B., Doel, T.R. and Brown, F. (1975), *J. gen. Virol.,* **29**, 189–198.
123. Schloemer, R.H. and Wagner, R.R. (1975), *J. Virol.,* **16**, 237–249.
124. Bishop, D.H.L., Repik, P., Obijeski, J.F., Moore, N.F. and Wagner, R.R. (1975), *J. Virol.,* **16**, 75–84.
125. Schoemer, R.H.and Wagner, R.R. (1975), *J. Virol.,* **15**, 882–893.
126. Stoffel, W., Anderson, R. and Stahl, J. (1975), *Hoppe-Seylers Z. physiol. Chem.,* **356**, 1123–1129.
127. Schlesinger, S., Gottlieb, C. Feil, P., Gelb, N. and Kornfeld, S. (1976), *J. Virol.,* **17**, 239–246.
128. Schlesinger, M.J., Schlesinger, S. and Burge, B. (1972), *Virology,* **47**, 539–541.
129. Duda, E. and Schlesinger, M.J. (1975), *J. Virol.,* **15**, 416–419.
130. Simons, K., Kaariainen, L., Renkonen, O., Chamberg, C.G., Caroff, H., Helenius, A., Keranen, S., Laine, R., Ranki, M., Soderlund, H. and Herman, G. (1973), In: *Membrane-Mediated Information,* (Kent, P., ed.), Vol. 2, Medical and Technical Publishing Co. Ltd., Lancaster, pp. 81–99.
131. Ranki, M., Kaarianinen, L., Renkonen, O. (1972), *Acta pathol. microbiol. scand. B.,* **80**, 760–768.
132. Garoff, H., Simons, K. and Renkonen, O. (1974), *Virology,* **61**, 493–504.
133. Sefton, B.M. and Keegstra, K. (1974), *J. Virol.,* **14**, 522–530.
134. Johnson, I. and Clamp, J.R. (1971), *Biochem. J.,* **123**, 739–745.
135. Keegstra, K., Sefton, B.M. and Burke, D. (1975), *J. Virol.,* **16**, 613–620.
136. Stollar, V., Stollar, B.D., Koo, R., Harrap, K.A. and Schlesinger, W. (1976), *Virology,* **69**, 104–115.

137. Utermann, G. and Simons, K. (1974), *J. mol. Biol.*, **85**, 569–587.
138. Kennedy, S.I.T. (1974), *J. gen. Virol.*, **23**, 129–143.
139. Meager, A., Ungkitchanukit, A., Nairn, R. and Hughes, R.C. (1975), *Nature*, **257**, 137–139.
140. Meager, A., Ungkitchanukit, A. and Hughes, R.C. (1976), *Biochem. J.*, **154**, 113–124.
141. Berge, T.O. (ed.) (1975), In: *International Catalogue of Arboviruses*, 2nd edn, U.S. Department of Health, Education and Welfare, Public Health Service, DHEW Publication No. (CDS) 75–8301, Washington, D.C.
142. Birdwell, C.R. and Strauss, J.H. (1974), *J. Virol.*, **14**, 672–678.
143. Mooney, J.J., Dalrymple, J.M., Alving, C.R. and Russell, P.K. (1975), *J. Virol.*, **15**, 225–231.
144. Nozima, T. and Yasui, K. (1970), *Acta Virol., Praha*, **15**, 174–.
145. Nozima, T., Yasui, K. and Homma, R. (1968), *Acta Virol., Praha*, **12**, 296–300.
146. Homma, R. (1968), *Acta Virol., Praha*, **12**, 385–396.
147. Yasui, K., Nozima, T., Homma, R. and Ueda, S. (1968), *Acta Virol., Praha*, **13**, 158.
148. Yasui, K., Nozima, T., Homma, R. and Ueda, S. (1971), *Acta Virol., Praha*, **15**, 7–18.
149. Andrewes, C.H., Bang, F.B. and Burnet, F.M. (1955), *Virology*, **1**, 176–184.
150. Waterson, A.P. (1962), *Nature*, **193**, 1163–1164.
151. Mahy, B.W.J. and Barry, R.D. (eds.) (1975), In: *Negative-strand Viruses*, Academic Press, London and New York, (a) Griffith, I.P. *ibid*, pp. 121–132. (b) Scheid, A. and Choppin, P.W. *ibid*, pp. 177–192. (c) Choppin, P.W., Lazarowith, S.G. and Goldberg, A.R. *ibid*, pp. 105–119. (d) Schulze, I.T. *ibid*, pp. 161–175.
152. Scheid, A. and Choppin, R.W. (1973), *J. Virol.*, **11**, 263–271.
153. Dourmashkin, R.R. and Tyrell, D.A. (1974), *J. gen. Virol.*, **24**, 129–141.
154. Huang, R.T.C., Rott, R. and Klenk, H.-D. (1973), *Z. Naturforsch.*, **28c**, 342–345.
155. Klenk, H.-D., Caliguiri, L.A. and Choppin, P.W. (1970), *Virology*, **42**, 473–481.
156. Lenard, J. and Compans, R.W. (1974), *Biochim. Biophys. Acta*, **344**, 51–94.
157. Fazekas De St. Groth, S. and Gottschalk, A. (1963), *Biochim. biophys. Acta*, **78**, 248–257.
158. Gottschalk, A. (1966), In: *Glycoproteins* (Gottschalk, A., ed.), Elsevier Publishing Co., Amsterdam, pp. 543–547.
159. Winzler, R.J. (1969), In: *Red Cell Membrane Structure and Function* (Jamieson, G.A. and Greenwalt, T.J., eds.), Lippincott, Philadelphia, pp. 157–171.
160. Tiffany, J.M. and Blough, H.A. (1971), *Virology*, **44**, 18–28.
161. Haywood, A.M. (1974), *J. mol. Biol.*, **83**, 427–436.

162. Woodruff, J.F. and Woodruff, J.J. (1974), *J. Immunol.*, **122**, 2176–2183.
163. Spear, P.G. and Roizman, B. (1972), *J. Virol.*, **9**, 143–159.
164. Pennington, T.H. (1974), *J. gen. Virol.*, **25**, 433–444.
165. Hochberg, E. and Becker, Y. (1968), *J. gen. Virol.*, **21**, 231–241.
166. Morgan, C., Rose, H.M. and Mednis, B. (1968), *J. Virol.*, **2**, 507–516.
167. Armstrong, J.A., Metz, D.H. and Young, M.R. (1973), *J. gen. Virol.*, **21**, 533–537.
168. De Larco, J. and Todaro, G.J. (1976), *Cell,* **8**, 365–371.
169. Segrest, J.P., Kahne, I., Jackson, R.L. and Marchesi, V.T. (1973), *Arch. Biochem. Biophys.*, **155**, 167–183.
170. Findlay, J.B.C. (1974), *J. biol. Chem.*, **249**, 4398–4403.
171. Winzler, R.J., Harris, E.D., Pekas, D.J., Johnson, C.A. and Webster, P. (1967), *Biochemistry,* **6**, 2195–2202.
172. Kornfeld, S. and Kornfeld, R. (1969), *Proc. natn. Acad. Sci., U.S.A.*, **63**, 1439–1446.
173. Morawiecki, A. (1964), *Biochim. biophys. Acta,* **83**, 339–347.
174. Marchesi, V.T., Tillack, T.W., Jackson, R.L., Segrest, J.P. and Scott, R.E. (1972), *Proc. natn. Acad. Sci., U.S.A.*, **69**, 1445–1449.
175. Uhlenbruck, G. (1964), *Vox. Sang.*, **9**, 377–384.
176. Marchesi, V.T. and Andrews, E.P. (1971), *Science,* **174**, 1247–1248.
177. Kathan, R.H., Riff, L.J.M. and Real, M. (1963), *Proc. Soc. exp. Biol. Med.*, **114**, 90–92.
178. Springer, G.F., Huprikar, S.V. and Tegtheyer, H. (1971), In: *Glycoproteins of Blood Cells and Plasma* (Jamieson, G.A. and Greenewalt, T.J., eds.), Lippincott, Philadelphia, pp. 35–49.
179. Suttajit, M. and Winzler, R.J. (1971), *J. biol. Chem.*, **246**, 3398–3404.
180. Peries, J.R. and Chany, C. (1962), *Proc. Soc. exp. biol. Med.*, **110**, 477–482.
181. Howe, C. and Lee, L.T. (1970), *Adv. Virus Res.*, **17**, 1–50.
182. Lerner, A.M. and Miranda, Q.R. (1968), *Virology,* **36**, 277–285.
183. Tillotson, J.R. and Lerner, A.M. (1966), *Proc. natn. Acad. Sci., U.S.A.*, **56**, 1143–1150.
184. Clamp, J.R. and Hough, L. (1965), *Biochem. J.*, **94**, 17–24.
185. Lukert, P.D. (1972), *Am. J. Vet. Res.*, **33**, 987–994.
186. Kodza, H. and Jungeblut, C.W. (1958), *J. Immunol.*, **81**, 76–81.
187. Verlinde, J.D. and De Baan, P. (1949), *Ann. Inst. Pasteur,* **77**, 632–641.
188. Haywood, A.M. (1975), *J. gen. Virol.*, **29**, 63–68.
189. Morgan, C., Rosenkranz, H.S. and Mednis, B. (1969), *J. Virol.*, **4**, 777–796.
190. Durand, D.P., Brotherton, T., Chalgren, S. and Collard, W. (1969), In: *The Biology of Large RNA Viruses* (Barry, R.D. and Mahy, B.W.J., eds.), Academic Press, New York, pp. 197–206.
191. Meiselman, N., Kohn, A. and Danon, D. (1967), *J. Cell Sci.*, **2**, 71–76.
192. Morgan, C. and Howe, C. (1968), *J. Virol.*, **2**, 1122–1132.
193. Chang, A. and Metz, D.H. (1976), *J. gen. Virol.*, **32**, 275–282.

194. Harrap, K.A. (1973), In: *Viruses and Invertebrates* (Gibbs, A.J., ed.), North Holland Publishing Co., Amsterdam, pp. 271–299.
195. Kawanishi, C.Y., Summers, M.D., Stoltz, D.B. and Arnott, H.J. (1972), *J. Invert. Path.*, **20**, 104–108.
196. Summers, M.D. (1969), *J. Virol.*, **4**, 188–190.
197. Summers, M.D. (1971), *J. Ultrastruct. Res.*, **35**, 606–625.
198. Knudson, D.L. and Tinsley, T.W. (1974), *J. Virol.*, **14**, 934–944.
199. Knudson, D.L. and Harrap, K.A. (1976), *J. Virol.*, **17**, 254–268.
200. Curtis, A.S.G. (1973), *Prog. Biophys. Mol. Biol.*, **27**, 317–386.
201. Hughes, R.C. (1975), *Essays in Biochemistry*, **11**, 1–36.
202. Lisowska, E. and Morawiecki, A. (1967), *Eur. J. Biochem.*, **3**, 237–241.
203. Harboe, A. (1963), *Acta path. microbiol. scand.*, **57**, 488–492.
204. Strandli, O., Mortensson-Egnund, K. and Narboe, A. (1964), *Acta path. microbiol. scand.*, **60**, 265–270.
205. Laver, W.G. and Webster, R.G. (1966), *Virology*, **30**, 104–115.
206. Marshall, R.D. (1974), *Biochem. Soc. Symp.*, **40**, 17–26.
207. Drzenik, R., Frank, H. and Rott, R. (1968), *Virology*, **36**, 703–707.
208. Laver, W.G. and Valentine, R.C. (1969), *Virology*, **38**, 105–119.
209. Nermut, M.V. and Frank, H. (1971), *J. gen. Virol.*, **10**, 37.
210. Padgett, B.L. and Walker, D.L. (1964), *J. Bact.*, **87**, 363–369.
211. Seto, J.T. and Rott, R. (1966), *Virology*, **30**, 731–737.
212. Janiel, R.I. and Kilbourne, E.D. (1966), *J. Bact.*, **92**, 1521–1526.
213. Webster, R.G. and Laver, W.G. (1967), *J. Immunol.*, **99**, 49–54.
214. Kilbourne, E.D., Laver, W.G., Schulman, J.L. and Webster, R.G. (1968), *J. Virol.*, **2**, 281–287.
215. Becht, H., Hammerling, U. and Rott, R. (1971), *Virology*, **46**, 337–341.
216. Seto, J.T. and Chang, E.S. (1969), *J. Virol.*, **4**, 58–66.
217. Springer, G.F. (1970), *Naturwessenshaft*, **57**, 162–171.
218. Pepper, D.J. (1964), *Biochim. biophys. Acta*, **156**, 317–326.
219. Gottschalk, A. and Fazekas De St. Groth, S. (1960), *Biochim. biophys. Acta*, **43**, 513–518.
220. Whitehead, P.H. and Winzler, R.J. (1968), *Arch. Biochem. Biophys.*, **126**, 457–464.
221. Morawiecki, A. and Lisowska, E. (1965), *Biochem. biophys. Res. Commun.*, **18**, 606–610.
222. Hughes, R.C. (1976), In: *Specificity of Plant Diseases*, (Wood, R.K.S. and Graniti, E., eds.), Plenum, New York-London, pp. 77–99.
223. Cuatrecasas, P. (1974), *Ann. Rev. Biochem.*, **43**, 169–214.
224. Cuatrecasas, P., Bennett, V., Craig, S., O'Keefe, E. and Sanyoun, N. (1976), In: *The Structural Basis of Membrane Function* (Hatefi, Y. and Diavadi-Chianiance, L., eds.), Academic Press, New York-London, pp. 275–291.
225. Thorne, H.V. (1973), *J. gen. Virol.*, **18**, 163–169.
226. Crittenden, L.B., Briles, W.F. and Stone, H.A. (1970), *Science*, **169**, 1324–1325.

227. Moldow, C.F., McGrath, M. and Van Santen, L. (1976), *J. Supramol. Struc.*, **4**, 497–506.
228. Rifkin, D.B. and Compans, R.W. (1971), *Virology*, **46**, 485–489.
229. Scheele, C.M. and Hanafusa, H. (1971), *Virology*, **45**, 401–407.
230. Dunnebacke, T.H., Levinthal, J.D. and Williams, R.C. (1969), *J. Virol.*, **4**, 505–513.
231. Ojinawi, N.K. and Olson, L.C. (1973), *Archiv. ges. Virusforsch*, **43**, 144–151.
232. Poste, G. (1970), *Adv. Virus Res.*, **16**, 303–356.
233. Lucy, J.A. (1970), *Nature*, **227**, 815–817.
234. Poste, G. and Allison, A.C. (1973), *Biochim. biophys. Acta*, **300**, 421–465.
235. Ahkong, Q.F., Fisher, D., Tampion, W. and Lucy, J.A. (1975), *Nature*, **253**, 194–195.
236. Haywood, A.M. (1974), *J. Mol. Biol.*, **87**, 625–628.
237. Miyamoto, K. and Gilden, R.V. (1971), *J. Virol.*, **1**, 395–406.
238. Swanson, J. (1975), In: *Microbiology – 1975*, (Schlessinger, D., ed.), Washington, American Society for Microbiology, pp. 124–126.
239. Poste, G. and Allison, A.C. (1973), *Biochim. biophys. Acta*, **300**, 421–465.
240. Hill, M.W. and Lester, R. (1972), *Biochim. biophys. Acta*, **282**, 18–30.
241. Beers, R.F. and Bassett, E.G. (eds.), (1976), *Cell Membrane Receptors for Viruses, Antigens, Antibodies, Polypeptide Hormones and Small Molecules*, 9th Miles International Symposium. Raven Press, New York.
242. Chan, V.F. and Black, F.L. (1970), *J. Virol.*, **5**, 309–312.
243. Boulanger, P.A. and Warocquier, R. (1972), *Exp. Mol. Pathol.*, **17**, 326–331.
244. Boulanger, P.A. and Hennache, B. (1973), *FEBS Letters*, **35**, 15–18.
245. Schlesinger, R.W. (1975), *Medical Biology*, **53**, 295–301.
246. Tillack, T.W., Scott, R.E. and Marchesi, V.T. (1972), *J. exp. Med.*, **135**, 1209–1227.
247. Couillin, M.M., Bou'e, A., Van-Cong, N., Weil, D., Rebourcet, R. and Fr'ezal, J. (1975), *C.R. Acad. Sci. [D] (Paris)*, **281**, 293–295.
248. Fiogen, K.J. (1975), *J. gen. Microbiol.*, **89**, 48–56.
249. Green, H. (1974), *N. Eng. J. Med.*, **290**, 1018–1019.
250. Inouye, S. and Norrby, E. (1973), *Archiv. Gesamte Virusforsch.*, **42**, 388–98.
251. Burrows, M.R., Sellwood, R. and Gibbons, R.A. (1976), *J. gen. Microbiol.*, **96**, 269–275.
252. Springer, G.F. and Desai, R.R. (1974), *Ann. Clin. Lab. Sci.*, **4**, 294–298.
253. Weiss, L. and Subjeck, J.R. (1974), *Int. J. Cancer*, **13**, 143–150.
254. Di Mosi, D.R., White, J.C., Schnaitman, C.A. and Bradbeer, C. (1973), *J. Bact.*, **115**, 506–513.
255. Hazelbauer, G.L. (1975), *J. Bact.*, **124**, 119–126.
256. Szmelman, S. and Hofnung, M. (1975), *J. Bact.*, **124**, 112–118.
257. Wayne, R. and Neilands, J.B. (1975), *J. Bact.*, **121**, 497–503.
258. Wayne, R., Frick, K. and Neilands, J.B. (1976), *J. Bact.*, **126**, 7–12.
259. Luckey, M., Wayne, R. and Neilands, J.B. (1975), *Biochem. biophys. Res. Commun.*, **64**, 687–693.

5

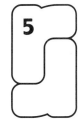

Acetylcholine Receptors

MOHYEE E. ELDEFRAWI

and

AMIRA T. ELDEFRAWI

Department of Pharmacology and Experimental Therapeutics
University of Maryland School of Medicine,
Baltimore, Md. 21201, U.S.A.

Abbreviations

ACh	– acetylcholine
HTX	– histrionicotoxin
H_{12}-HTX	– perhydrohistrionicotoxin
α-BGT	– α-bungarotoxin
QNB	– quinuclidinyl benzilate
EPSP	– excitatory postsynaptic potential
IPSP	– inhibitory postsynaptic potential
EPP	– endplate potential
MEPP	– miniature endplate potential
ICM	– ion conductance modulator
cGMP	– guanosine 3′: 5′-cyclic monophosphate
SDS	– sodium dodecylsulfate

Acknowledgments

We are grateful to Dr Edson X. Albuquerque for his critical review of, and helpful comments on, this chapter, to Dr. Joseph Byron for his valuable suggestions on Section 5.7.3 and Dr John Rash for kindly providing Fig. 5.2b.

Our research on ACh receptors was supported by National Science Foundation Grant BNS76–21683 and National Institutes of Health Grants NS–13231 and AI–13640.

Receptors and Recognition, Series A, Volume 4
Edited by P. Cuatrecasas and M.F. Greaves
Published in 1977 by Chapman and Hall, 11 New Fetter Lane, London EC4P 4EE

5.1 INTRODUCTION

Communication between cells in our bodies is mainly by means of chemical messengers. These chemicals or hormones may be simply classified into two broad categories: long-range messengers and short-range ones. The former are synthesized by specialized glands, released into the blood and circulated intact to their target organs. Examples are insulin and thyroid. On the other hand, the short-range messengers are small chemicals produced by nerve cells (neurons) and are released only a few angstroms or microns away from their target cells. They are called neurotransmitters, and an example is acetylcholine (ACh). There are also intermediate range messengers, we may call modulators, which are produced by one of several different cells. Chemical messengers are recognized and bound by highly selective and stereospecific sites on molecules which are present in the target cells. These molecules are called receptors.

The vertebrate body contains at least three types of ACh receptor which differ in their action and in their pharmacology. These receptors influence cellular responses (e.g. muscle contraction, glandular excretion) either directly or indirectly through the action of second messengers.

The study of the effects of specific poisons slowly led to the discovery of ACh transmission. Starting with Claude Bernard's classic experiments, described in 1857, which led to his conclusion that *d*-tubocurarine (a specific ACh receptor inhibitor) acted specifically and inhibited the skeletal nerve-muscle junction [1]. After many years of further study, it was suggested that the effect of nicotine on ganglion transmission was on the postsynapse [2] and in 1906 Langley proposed the term 'receptive substance' for what we now call ACh receptor [3]. The first discovery of the action of ACh came later in 1914, when Dale [4] tested the actions of many esters and ethers of choline. That ACh was a neurotransmitter became known from Loewi's experiments in the twenties when a perfusate from a vagal-stimulated frog heart slowed down the heart of another [5], and this so-called 'Vagusstuoff' was suggested to contain ACh [6].

It is now hard to believe that, despite the wealth of information physiologists and pharmacologists gathered on cholinergic synaptic transmission (synapses where ACh is the transmitter), all we knew about the ACh receptor until the late sixties was that it should exist. The

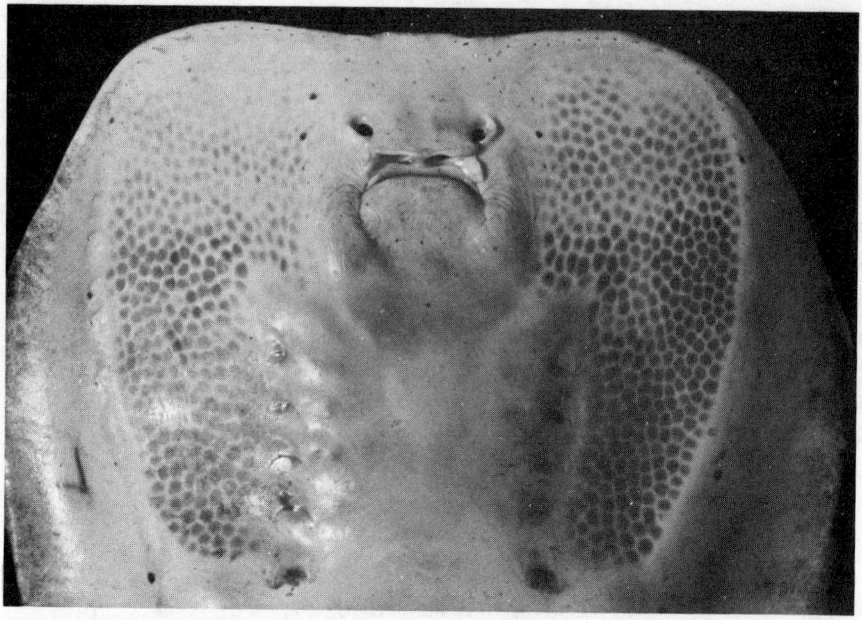

Fig. 5.1 Ventral view of *Torpedo ocellata.* The bilateral kidney-shaped honeycomblike structure is the electric organ (from Eldefrawi [80]).

major difficulty was a suspicion that the functions of ACh receptors (e.g. changes in membrane potential, muscle contraction) could be detected only in intact preparations. However, the studies of Sutherland, Rall and their colleagues on hormone receptors and second messengers in the fifties [7, 8] dispelled the belief that hormone recognition and cellular effects required an intact cell and proved that hormone action could indeed be studied in broken cell preparations. Thus started a new era in receptor research.

Interest in the production of electric current by electric organs of fish dates back to Faraday's studies on the electric eel, *Electrophorus electricus,* in 1758 [9] and Walsh's studies on the electric ray, *Torpedo* (Fig. 5.1), in 1773 [10]. Electric organs of fish were discovered to be tissues that were very rich in ACh and ACh-esterase (an enzyme that hydrolyzes ACh and thus helps in terminating the action of ACh) [11–13], and accordingly possibly the ACh receptors. Extensive electrophysiologic studies were conducted on the single cell of an electric organ cell (called electroplax), and data were accumulated on the pharmacology of their ACh receptors [14]. Embryonically, they were found to be derived from skeletal muscle, and the pharmacology of their ACh receptors was identical with those in such muscles [12]. While they have lost their contractile ability they developed an ability to generate electric potential. The relatively ready availability of these tissues, their content of only one kind of neurotransmission, their richness in ACh receptors (amounting to 1% of all the proteins in the *Torpedo* electric organ) and the accumulated data on their electrophysiology and pharmacology are the reasons why the ACh receptor is the first neurotransmitter receptor to be isolated in purified form. Research on these receptors has helped greatly in developing the new technology required for and utilized in the isolation of other neurotransmitter and hormone receptors.

Because the status of our knowledge of the biochemical nature of the nicotinic ACh receptors is far more advanced than that of muscarinic ACh receptors, and of electric organ and neuromuscular receptors more so than central ACh receptors, emphasis in this chapter shall be on skeletal neuromuscular and electric organ ACh receptors. Comparisons with the other ACh receptors shall be made whenever appropriate. The term 'ligand' shall be used throughout this chapter to denote any molecule (e.g. ACh, drug or toxin) that interacts with the ACh receptor.

5.2 GENERAL CHARACTERISTICS OF ACh RECEPTORS

5.2.1 Distribution and physiologic function

There are several kinds of synaptic junctions, those between neuron and neuron, neuron and muscle or neuron and gland (Fig. 5.2). A neuron receives many synapses along its axon, dendrites or cell body. In smooth muscles, most invertebrate muscles [15] and the frog tonic skeletal muscles the motor nerve branches terminate at many points along the entire length of the muscle fiber, and the entire fiber surface is sensitive to ACh [16]; thus ACh receptors are distributed all over the fiber. The skeletal neuromuscular junction is highly specialized and organized with a postsynaptic membrane that is greatly folded, only the tops of the folds containing ACh receptors (20 000–40 000 μm^{-2} [17, 18]. Each muscle fiber receives single innervation.

In cholinergic neurons (e.g. the vertebrate motoneuron) ACh is synthesized and stored in the nerve terminal in synaptic vesicles. A nerve impulse causes release of ACh [19], which binds to its receptors on the postsynaptic membrane causing a localized graded response. There are

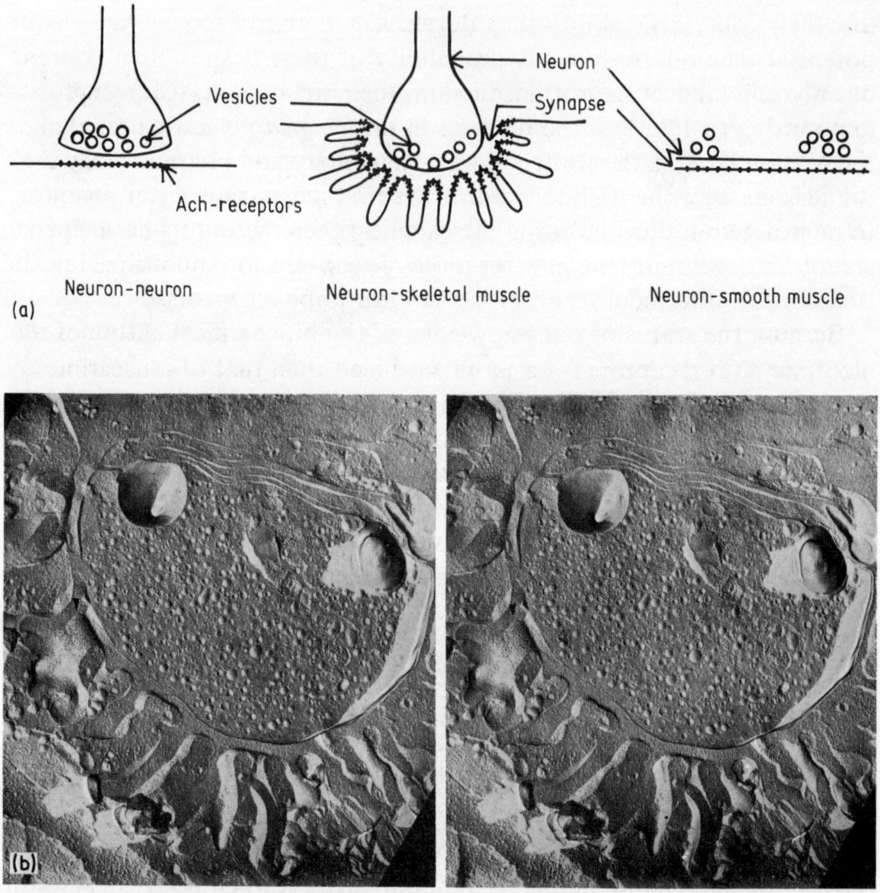

Fig. 5.2 (a) Diagrammatic representation of different cholinergic synapses. (b) Stereo electronmicrograph of freeze-fractured endplate of a rat extensor digitorum longus muscle (17 500 x magnification) (courtesy of Dr John Rash).

several types of cellular responses that are mediated by ACh receptors:
(1) Excitatory; where the normally polarized membrane (positive on the outside and negative on the inside) becomes depolarized as a result of the inward flux of Na^+ and outward flux of K^+. In a neuron, the localized depolarization is called excitatory postsynaptic potential (EPSP) and in a skeletal muscle it is an endplate potential (EPP).
(2) Inhibitory; where the polarized membrane becomes hyperpolarized as a result of the outward flux of K^+, inward flux of Cl^- or both. This is

called inhibitory postsynaptic potential (IPSP). The IPSP opposes the effect of EPSP at the same site and time. If the algebraic summation of these potentials exceeds a certain threshold a wave of depolarization, called an action potential, is propagated in a neuron, most skeletal muscle and cardiac muscle. In most smooth muscle, certain tonic skeletal muscle and gland cells, there is no propagated impulse; instead there is local depolarization and slow contraction of the muscle or a gland secretion. There may be other types of postsynaptic cholinergic response such as that found in the ganglia of the marine mollusc *Aplysia,* a diphasic potential response, an initial depolarization, due to increased Na^+ conductances, followed by a small late inhibitory component due to increased K^+ conductance [20]. These different responses may be due to differences in the ACh receptors (see Section 5.2.2) and/or the ion conductance modulator (ICM), associated with the receptor in each of these types of synapses (see Section 5.5.4).

The EPP is caused by the release into the junction of several hundred packets (quanta of about 10 000 molecules each) of ACh from the nerve ending by the arrival of an impulse. Miniature endplate potentials (MEPP) are caused by the spontaneous release of a single quantum of ACh in the absence of any form of stimulation [21].

How does binding of ACh to its receptor activate the ICM and cause ion transport? Katz and Miledi [22–25] suggest that, when ACh is released in a cholinergic synapse, there occur high-frequency collisions between ACh molecules and ACh receptors and also that each collision or several collisions open a single ionic channel for an average duration of 1 millisecond causing a depolarization equivalent to 0.22 μV at 22°C. Various receptor activators (or agonists) cause the ionic channel to open for different durations, ranging from 0.12 to 0.35 ms for acetylthiocholine, decamethonium and carbamylcholine as compared to 1 ms for ACh and 1.65 ms for suberyldicholine at 22°C [26]. The transient response seen at cholinergic synapses is due mainly to removal of ACh by the hydrolytic activity of the enzyme ACh-esterase [27]. This enzyme, which is abundant in cholinergic synapses, thus serves to terminate the action of ACh on the receptor. Inhibition of this enzyme by specific drugs would be expected to increase the time of reaction between ACh and the receptor and it does.

Stevens and collaborators [28–31] suggest that binding of ACh to its receptor reduces the energy barrier for a conformational change in the ACh receptor (see Section 5.5.1) and possibly in associated channel macromolecule(s). In the endplate, this causes the channel conductance to increase from effectively zero to 2 to 3 x 10^{-11} mho. The concentration of ACh in the synaptic gap decreases rapidly through hydrolysis by ACh-esterase (which is abundant in the gap) and reabsorption of the resultant choline as well as through diffusional losses. Since the closing of the opened channels is relatively slow (about 1 ms), the concentration of ACh drops essentially to zero before all the endplate channels close. Thus, the relaxation of the gating molecules to their closed conformation, and not ACh concentration, is the rate-limiting step in endplate conductance changes. Opening and closing of the channels is voltage-dependent and, in the process, the dipole moment of the molecule is altered by 50 Debye or more in the electric field produced by the membrane potential. The energy of the channel molecule in a certain conformation is dependent upon the membrane potential. Therefore, the opening and closing rates are influenced by membrane potentials. They also propose that the receptors with bound ACh are continually opening and closing channels in a probabilistic manner, while the mean number of open channels is fixed [31].

Muscarinic responses (see Section 5.2.2) differ from nicotinic ones in having a much longer latency ($\simeq$ 100 ms compared to 0.22 ms), and the duration after a stimulus is much longer (500 ms or longer compared to, at most, 30–100 ms) [16, 32, 33]. This difference undoubtedly reflects an ICM of a different nature, possibly involving an enzyme reaction (nucleotide cyclase) and the recruiting of a second messenger intracellularly (see Section 5.5.5).

5.2.2 Pharmacologic classification

ACh receptors of the peripheral nervous system of vertebrates are classified into two major categories on the basis of their sensitivity to a variety of agonists (drugs whose binding to the ACh receptor mimics the effect of ACh) and antagonists (drugs whose binding to the ACh receptor inhibits the effect of ACh) (Fig. 5.3). There are *nicotinic* receptors, activated by nicotine and inhibited by *d*-tubocurarine, which are found in autonomic ganglia, skeletal muscle and in the central nervous system and there are *muscarinic* receptors, activated by L(+) muscarine and pilocarpine and inhibited by atropine and quinuclidinyl benzilate (QNB) (Fig. 5.4), which are found in smooth muscles, cardiac muscles, ganglia, brain and glands. There are also a few ACh receptors in mammalian brain that have been found to exhibit mixed nicotinic-muscarinic pharmacology [34, 35].

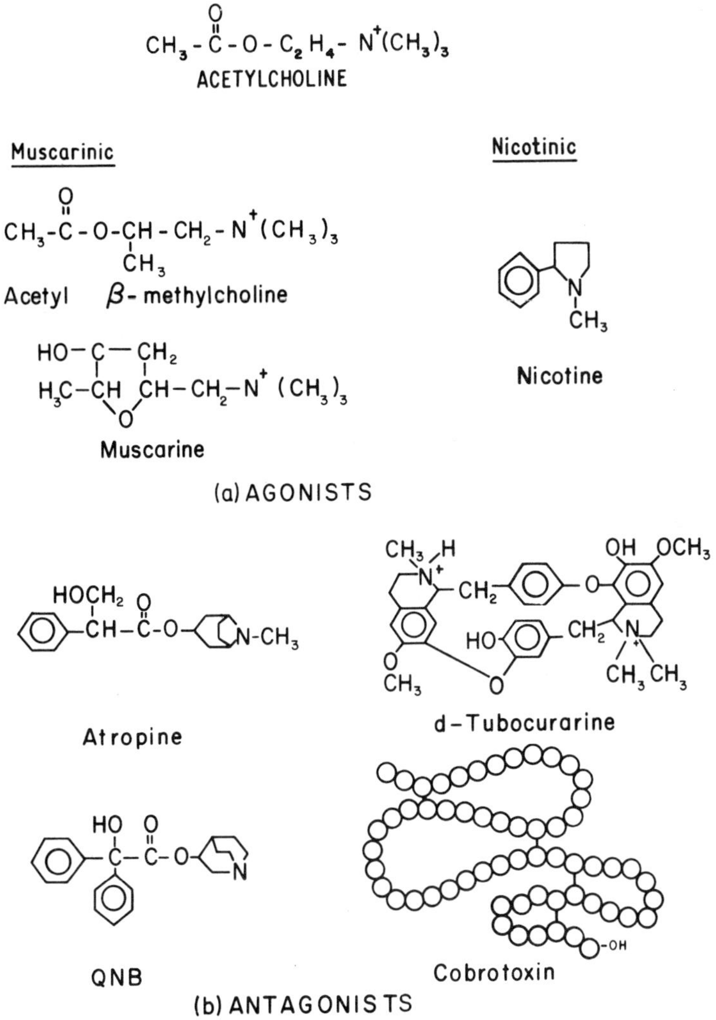

Fig. 5.3 Acetylcholine, some receptor agonists (a) and antagonists (b). Muscarinic drugs are on the left side and nicotinic ones on the right. Each amino acid in cobrotoxin is represented by a circle.

This pharmacologic classification reflects differences in the recognition sites between nicotinic and muscarinic receptors. However, despite these differences, ACh binds to both. Studying the crystal structure of ACh and various drugs, Pauling, Chothia and co-workers [36–38] suggested that different groups of the ACh molecule bind with each type of

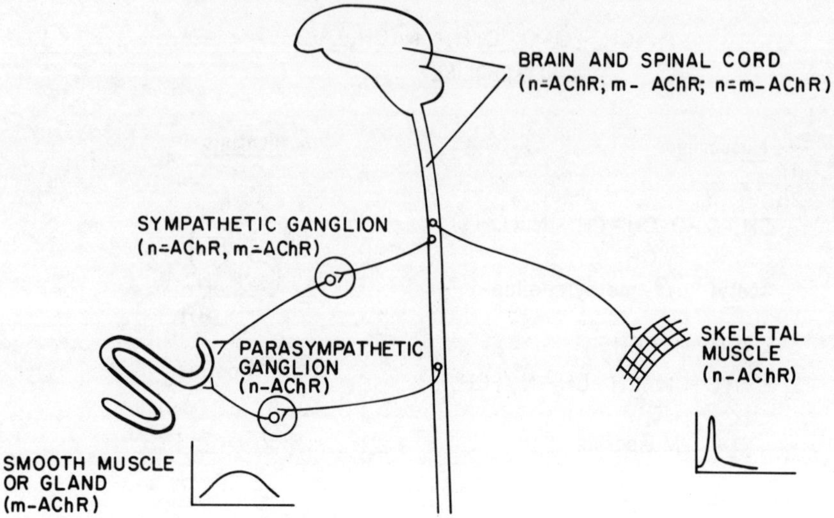

Fig. 5.4 Schematic representation of the distribution of nicotinic ACh receptors (*n*-AChR), muscarinic ACh receptors (*m*-AChR) and mixed type receptors (*n*-*m*-AChR) in the central and peripheral nervous tissue. The muscle response elicited by nerve stimulation is illustrated in the lower left diagram in the case of smooth muscle and lower right diagram in the case of skeletal muscle.

receptor (nicotinic or muscarinic). Studies on ACh analogs showed that a slight variation in structure changes the molecule from strongly nicotinic to strongly muscarinic or vice versa. An example is L(+)-acetyl-α-methylcholine for the former and L(+)-acetyl-β-methylcholine for the latter. Muscarone (Fig. 5.5) is an interesting molecule because its structure resembles that of ACh. It is not an ester, and thus is not hydrolyzed by ACh-esterase; consequently it was the first ligand to be successfully used for the *in vitro* identification of ACh receptors. It is also a nicotinic as well as a muscarinic agonist [39]. The potent muscarinic agonists usually have the L(+) configuration, which is 200 times more active than the D(−). Muscarone is an exception, the D(−) form is the more potent one and by only three-fold [39]. Study of its conformation in crystals led to the suggestion that this is due to the N^+-C-C-O torsion angle and the ring conformation [38].

The finding that hexamethonium is an effective antagonist of the nicotinic ACh receptors of mammalian ganglia but not of those of skeletal neuromuscular ones, while decamethonium is more effective on

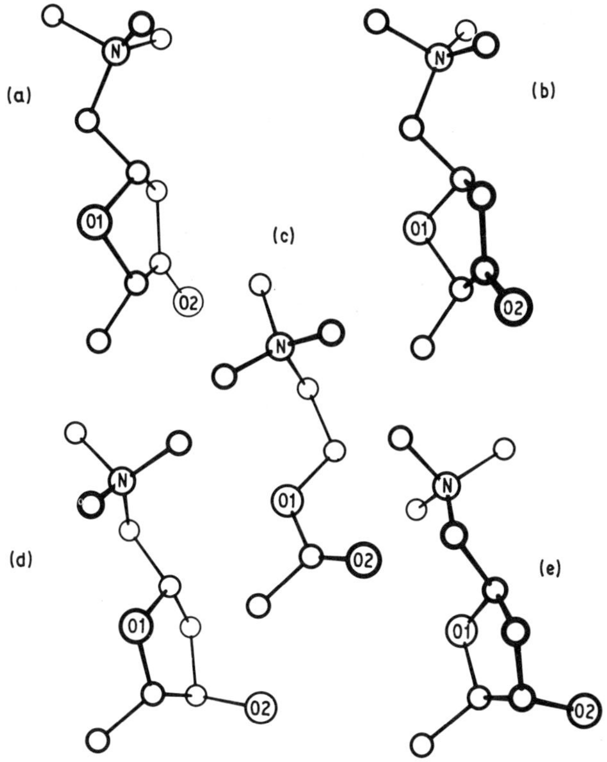

Fig. 5.5 Drawings of the enantiomers of muscarone, muscarine and ACh in the conformation relevant to the muscarinic receptor. (a) L(+)-muscarone; (b) D(−)-muscarone; (c) ACh; (d) L(+)-muscarine; (e) D(−)-muscarine (from Pauling and Petcher [38]).

the latter type, indicates that there are subclasses of nicotinic receptors. Some drugs, such as decamethonium, activate the ACh receptor then inhibit it in man, and thus are called depolarizing blockers. In contrast, *d*-tubocurarine blocks the receptor in neuromuscular junctions without causing depolarization and so do most antagonists.

The most specific nicotinic compounds are components of snake venoms, such as the α-neurotoxin of the elapid snake, *Bungarus multicinctus,* and the α-neurotoxin or cobrotoxin of cobra venoms. Because of their high-affinity binding to skeletal muscle and electric organ ACh receptors, these toxin played a major role in the *in vitro* identification of ACh receptors and in their purification. They have also been used to identify brain and ganglionic ACh receptors [40–42], although it is not yet certain that they have the same specificity for these receptors. Very recently, Lee and Chen (personal communication) reported that rats injected intravenously with cobrotoxin died by peripheral respiratory paralysis while their phrenic vagal nerve was still active, suggesting that by the time α-BGT inhibited neuromuscular receptors, it had not inhibited (or maybe did not reach) some brain nicotinic ACh receptors. On the other hand, surugatoxin, a toxin extracted from the Japanese ivory mollusc, *Babylonia japonica*, has very recently been found to antagonize the depolarizing action of carbamylcholine in rat superior cervical ganglion, and was suggested to be selective for ganglionic nicotinic ACh receptors [43].

ACh receptors are also found in insects. Neuromuscular transmission is not cholinergic in insects [44, see also 45], but ACh is a transmitter in the insect central nervous system [46, 47, see also 15, 45]. Nicotinic and muscarinic receptors have been detected electrophysiologically in the cockroach sixth abdominal ganglion by the Roeders [48, 49] and Shankland and co-workers [50, 51]. Recent biochemical and electrophysiological studies on insect brain suggest that most of its ACh receptors have a mixed nicotinicmuscarinic character [52–54, and D. Sattelle, personal communication] (Table 5.1). These receptors, however, are apparently unaffected by the snake neurotoxins even though they recognize nicotinic drugs (e.g. nicotine, *d*-tubocurarine and decamethonium). It may be that this mixed pharmacology is characteristic of more primitive ACh receptors, which are found in insect brain and in a few cases in mammalian brain stem.

5.2.3 Density and relationship to response

In the highly specialized endplate of skeletal muscles, densities of ACh receptors have been calculated by use of the specific radiolabeled α-BGT *in vitro* or *in vivo.* For a given muscle fiber type and size, the endplate has a fixed number of receptors ($10^7 - 10^8$), which is greater in larger fibers and when the postsynaptic folding is increased [55].

The most accurate method of determining the extent of membrane depolarization evoked by their reaction with a given quantity of released ACh is focal application of ACh directly on the endplate using an appropriate optical system (e.g. Nomarski) [17]. ACh sensitivity, which is expressed as the depolarization produced per amount of ACh released by current from a micropipette, has been determined by

Albuquerque and co-workers [17, 55] to range from 5000 to 7000 mV per nanocoulomb for all muscle types and species studied. Thus, for every million ACh molecules applied, 3.5 mV is obtained, only 3–16% of that calculated for ACh released by vesicles. The discrepancy may be due to faster loss of applied ACh by diffusion and the maximized geometric advantage obtained for ACh released at the right point by vesicles and its restricted diffusion within the fold. It has been calculated that the released ACh molecules diffuse in the junction during the 300 μs period of endplate current generation, and cover an area of postsynaptic membrane of about 3 μm^2 [55, 56]. Since the same sensitivity was obtained for endplates with different receptor numbers, it was concluded that the sensitivity was determined by the local density of ACh receptors in the crest of the synaptic folds (28 000–40 000 μm^2) [55].

If the number of ACh receptor sites in an endplate (6×10^7) is compared to the number of ACh molecules released in one impulse (6×10^6) [57, 58] we find that there is a 10-fold excess of receptor sites. Some of the ACh released is probably hydrolyzed before it reaches the ACh receptor, but most of the molecules that reach the receptors bind to them. When ACh-esterase in frog endplates was inhibited, 66% of the ACh molecules released bound to the receptors [59]. Albuquerque, Barnard and co-workers [60] studied the relationship between ACh-receptor blockade by [^{3}H]α-BGT and percentage inhibition of EPP or muscle twitch response in rat skeletal muscle. Their results indicated that up to 25% of receptors could be blocked without inhibiting the response and these may represent a reserve of 'spare receptors'.

Other effector cells (e.g. smooth muscles, glands and nerve cells) which carry ACh receptors have not been studied as extensively as the skeletal muscles, thus density–response relationships are still unclear. Furthermore, there are effector cells in the body carrying ACh receptors, which respond to unbelievably low concentrations of ACh (see Section 5.7.3), yet we know nothing of their receptor densities or even the immediate cellular response controlled by these ACh receptors.

5.3 METHODS OF STUDYING ACh RECEPTORS

5.3.1 Cellular responses

Until nine years ago, all studies of ACh receptors were indirect, drawing inferences from the cellular response to receptor activation or inhibition. The effects of drugs on cholinergic transmission are studied by a variety of methods such as measurement of muscle twitch or contraction, membrane potential, ionic flux and composition, ultramicrosopic changes and drug uptake. Certain preparations and experimental approaches became widely used for studying specific types of ACh receptors, examples of which are presented below.

Mechanical

The simplest method used to study cellular responses triggered by ACh receptors is the recording of a mechanical response, namely muscle contraction. A widely used preparation is the guinea-pig ileum, an example of which is that used by Triggle and Triggle [61]. A segment ($\simeq$ 2 cm) of the longitudinal muscle is separated from the underlying circular muscle and suspended in a jacketed glass bath of 10 ml of aerated buffer. The lower end of the muscle is fixed while the top end is connected to a light lever, and the mechanical responses are recorded on smoked paper. Various concentrations of different drugs are applied to the bath and the effect on muscle contraction recorded.

Electrical

The introduction of intracellular microelectrode technology in 1945–1950 [21, 62, 63] and the development of several neuromuscular preparations easily obtainable from a number of vertebrate species has given us a wealth of information about cholinergic transmission in general and ACh receptors of skeletal muscles in particular. Intracellular recordings of endplate potential depend on the use of microelectrodes with tips small enough to penetrate cell membranes without damaging them and resulting in deterioration of the cell. These small tips give high resistance of the order of 5–7 ohms so that it is necessary to use a device of high input impedance and low output resistance between the electrode and recording apparatus. Such a device was called cathode follower and nowadays is often described as an operational amplifier. By stimulating the nerve to the muscle, it causes a resultant depolarization of the surface membrane which lasts several tens of milliseconds. It reaches a peak at about 1 millisecond and declines to a half in another few milliseconds. This is consistent with the idea that the endplate potential is generated by a rapid displacement of charge across the postsynaptic membrane. The other method used to study events occurring at the postsynaptic membrane when the transmitter reacts with the recognition site is the voltage clamp technique. In this case, the membrane potential is kept constant by an electronic feedback circuit. Another method is simply recording an endplate potential by extracellular recordings, thus allowing the appearance of endplate current at that particular membrane potential. Microelectrode technology also made it possible to study chemical transmission in central tissue. Prior to the introduction of biochemical studies, most of our knowledge of ACh receptors of mammalian brain or ganglionic ACh receptors, their localization and their pharmacology, was obtained by electrical measurements.

The development of the monocellular electroplax preparation by Schoffeniels and Nachmansohn in 1957 [14, 63] was a milestone in studies on the physiology and pharmacology of the ACh receptor in *Electrophorus* electroplax. The method is feasible because of the large size of a single cell from the posterior part (the bundle of Sachs) of the *Electrophorus* electric organ (1 mm in cross-section and 5–10 mm in length). The chamber used is made of two blocks of lucite each with a pool cut out and, when held together by a pin, the two pools form a single one separated by a sheet of nylon having a window, the electroplax and a grid of nylon threads. Compounds dissolved in one pool could pass to the other pool only through the cell and there is no leakage. Either extracellular electrodes are used, one stimulating and one recording electrode, placed in each of the chambers; or an intracellular microelectrode is inserted into the electroplax and another in one of the solutions.

A similar preparation from *Torpedo* electroplax was tried, but the thinness and fragility of this electroplax rendered it impossible to use in the above experimental conditions. However, intracellular recordings of membrane potential of *Torpedo* electroplax have recently been made [64].

Biochemical

If activation of ACh receptors on a certain cell lead to a discrete biochemical reaction (e.g. activation of an enzymic reaction) then this enzymic reaction may be monitored as an index of receptor activity.

There is strong evidence that the activation of muscarinic receptors in some effector cells leads to increased levels of cGMP (see Section 5.5.1). Thus, one method of studying these ACh receptors in cells relies on increased production of cGMP by activation of guanyl cyclase and phosphorylation of membrane proteins. In one study [65], slices of the brain were suspended in buffer and gently agitated, then after various periods, the supernatant obtained from the slices and the incubation medium assayed for cGMP by radioimmunoassay [66] or by co-chromatography with [^{3}H] cGMP on silica gel thin-layer chromatographic plates [67]. Thus, presence of ACh receptors and their activation or inhibition by drugs can be detected by this cellular response.

5.3.2 ACh receptors in broken cell preparations

Unlike enzymes whose presence is easily detectable in broken cell preparations by their catalytic reaction products, receptors can be detected in such systems only by their stoichiometric binding of radiolabeled agonists, antagonists or modifying agents, specific antibodies, or by use of specific fluorescent probes and the monitoring of changes in fluorescence caused by binding of various drugs. The following example illustrates the difficulty in working with stoichiometric receptor-binding reactions. If we were to measure the concentration of ACh-esterase and ACh receptors in tissue homogenate containing 10^3 active sites of each of the enzyme and receptor, we find that the enzyme (E) reacts with ACh according to equation 5.1, while the reaction of the receptor (R) with ACh is described by equation 5.2.

$$E + ACh \leftrightarrow [E\text{-}ACh] \rightarrow E\text{-acetyl} \rightarrow E + \text{acetate} \quad \searrow + \text{choline} \qquad (5.1)$$

$$R + ACh \leftrightarrow [R\text{-}ACh] \qquad (5.2)$$

Assuming that ACh carries a [^{3}H] label on the acetate moiety, and knowing that the enzyme has a turnover number of 7.5×10^5 molecules/min/catalytic site, we anticipate finding ($7.5 \times 10^5 \times 10 \times 10^3$) 7.5×10^9 molecules of [^{3}H] acetate after 10 minutes of reaction time. On the other hand, at saturation, 10^3 receptor sites can bind only 10^3 ACh molecules (one/active site). In other words, the catalytic activity of a few enzyme molecules amplifies the reaction to a measurable level, while the stoichiometric reaction of a few receptors with substrates is impossible to measure. For that reason, very high specific activity isotope labelling of receptor ligand is usually needed.

An important principle to remember when studying binding of ligands to putative receptors in subcellular preparations is that 'everything binds to everything'. Most drugs (particularly those positively charged) bind to most proteins if a high enough concentration of the drug is used. Thus, it is binding of a high affinity that one looks for, one that is inhibited by the drugs and toxins that bind to the receptor *in vivo*. Active and inactive stereoisomers are very useful in identifying the

Table 5.1 Drug profiles of four different ACh receptors. Binding of the nicotinic receptors (e.g. in *Torpedo* electroplax and muscle cell culture) is blocked largely by nicotinic drugs (*d*-tubocurarine and nicotine), while binding of the muscarinic receptors of cow brain is blocked largely by muscarinic drugs (atropine and pilocarpine). Binding of the mixed nicotinic–muscarinic type receptors of housefly brain is blocked by both nicotinic and muscarinic drugs.

Drug	% Blockade of binding			
	Torpedo electroplax* $[^3H]$ ACh (0.1 μM)	Muscle cell culture† $[^{125}I]$ α-BGT (1.25 nM)	Cow brain ‡ $[^3H]$ QNB (1 nM)	Housefly brain § $[^3H]$ decamethonium (0.1 μM)
d-Tubocurarine	60	75	9	40
Nicotine	41	95	26	50
Atropine	10	6	100	80
Pilocarpine	12	0	54	84

* Data from Eldefrawi *et al.* [93] on Lubrol-solubilized electric organ membranes. Blocking drugs were used at 10 μM.

† Data from Vogel *et al.* [236] on cultured chick skeletal muscle. Blocking drugs were used at 10 μM.

‡ Unpublished data from Eldefrawi and Schuessler on membranes of corpus striatum. Blocking drugs used at 10 μM except for atropine, which was used at 1 μM.

§ Data from Eldefrawi *et al.* [112] on the partially purified putative ACh receptors. Blocking drugs were used at 10 μM.

receptor *in vitro* (see Table 5.1).

The various methods used in studies of ACh receptors in broken cell preparations rely either on separation of receptor–ligand complex from free ligand as in filter assays or gel filtration, on differences in concentration of free ligand before and after exposure to receptors as in centrifugal assay, or of free ligand from the additive concentration of free ligand plus receptor-bound ligand as in equilibrium dialysis.

The validity of the *in vitro* methods for identification of ACh receptors was established when the same concentrations of receptor sites were obtained by different techniques and different ligands. The ACh receptor concentration of *Torpedo marmorata* electric organ was calculated to be 1 nmol/g tissue, whether by use of equilibrium dialysis of reversibly binding ligands [68, 69] or by the binding of [^{131}I]α-BGT determined by Miledi *et al.* [70] by means of centrifugation and gel filtration. The concentration of ACh receptors in the electric organ of *Electrophorus electricus* was calculated to be 0.02–0.03 nmol/g tissue by the affinity label reagent *N*-ethylmaleimide [71] and by equilibrium dialysis of reversibly binding cholinergic ligands [72, 73] or filter assay of α-BGT binding [74]. In mammalian brain, the concentration of muscarinic ACh receptors was determined to be 3 pmol/g mouse brain by equilibrium dialysis of reversibly binding ligands [75] (equal to 302 pmol/g protein) and 400 pmol/g protein of rat cortex by the irreversible binding of *N*-2′-chloroethyl-*N*-[2″, 3–$^3H_2$]propyl-2-aminoethyl benzilate [76]. Nicotinic receptors were 3 pmol/g mouse brain as determined by equilibrium dialysis [75] and 3.4 or 2.1 pmol/g rat and guinea-pig brains respectively, using α-BGT binding [77].

Centrifugation

This was the first method to be used to identify ACh receptors *in vitro* and was introduced by O'Brien and Gilmour in 1969 [78] for binding of reversible ligands. Membranes, prepared from the tissue, are incubated with the radiolabeled ligand and after up to 1 h, the mixture is centrifuged (11 000 x *g* for 15 min) and the radioactivity of samples from the supernatant counted. The reduction in radioactivity of this sample compared with that of one taken before exposure to the membranes represents the amount of ligand bound to the membranes. Although this method correctly identified the *Torpedo* ACh receptor, it has a low sensitivity when a reversibly binding ligand is used, since it relies on a small difference between two large numbers and some radiolabeled ligand may be trapped within, or lightly adsorbed to, the pellet. Centrifugation has been quite successful when used for studies of the quasi-irreversible binding of α-BGT [70] and particularly sucrose density-gradient centrifugation [70, 79].

Equilibrium dialysis

In many cases, the study of interactions of reversibly binding ligands with the ACh receptor has utilized equilibrium dialysis [80]. In this method, a small volume (e.g. 0.3–1 ml) of the membrane suspension or solution under study is placed in a dialysis bag tied at both ends. The bag is made of semi-permeable membrane such as cellophane, so that the protein and other large molecules are retained inside whereas salts and radiolabeled drugs pass through it freely. The bag is suspended in a large volume of high ionic strength buffer containing a known concentration of radiolabeled ligand. The flask is shaken until equilibrium of the ligand is reached, and if the ligand binds to the receptor the radioactivity inside the bag would be higher than that in an equal sample from the bath. The effect of a drug on the binding of the receptor is studied by placing it in the dialysis bath with the radiolabeled ligand. If the drug (agonist or antagonist) binds to the receptor, this results in reduction of the binding of the specific radiolabeled ligand. This method can be used for membrane-bound, solubilized or purified ACh receptors and is particularly suited for tissues that are rich in ACh receptors.

In a membrane where the ACh receptor constitutes only a small component of the membrane, it is necessary in equilibrium dialysis experiments to correct for the volume occupied by the membrane [81]. This was an important factor in studies on *Electrophorus* ACh receptors, but may not be so in *Torpedo* receptors, whose electric organ has 20–50 fold more ACh receptors per g tissue,

Filter assay

This method is particularly suited for the quasi-irreversible binding of α-BGT. It was used by Franklin and Potter [82] for studies on membrane-bound and solubilized *Torpedo* receptors. The diluted membrane preparation in buffer is incubated with [^{131}I]α-BGT for varying periods, then 0.5 ml aliquots are diluted into 10 ml solution containing bovine serum albumin (to reduce non-specific adsorption of toxin to filters) and filtered on 0.45 μm Millipore HAWP filters. The filters are washed with the same solution then counted for radioactivity. For Triton-solubilized ACh receptors the toxin receptor complex is separated from free toxin by its precipitation with 33% saturated ammonium sulfate, pH 5.

The sensitivity of the method is low because the amount of α-BGT that binds to the membrane is relatively high in comparison with that binding to ACh receptors in the preparation.

The sensitivity of filter assay was improved by Schmidt and Raftery [83] who used DEAE cellulose filters, which are positively charged. Since α-BGT is positively charged at pH 7, while the ACh receptor and its complex with α-BGT are negatively charged, little free α-BGT binds to these filters. Glass fiber filters (e.g. Whatman GF/C) are also used for membrane-bound ACh receptors. Their non-specific binding is low and thus they are particularly useful for reversibly binding ligands. These filters are used with excellent results for binding of [^{3}H]QNB to membrane-bound muscarinic receptors of mammalian brain and also of [^{125}I]α-BGT to membrane-bound *Torpedo* (Eldefrawi, unpublished). Non-specific binding to glass is reduced by the pre-application of unlabeled α-BGT before filtration of receptor–α-BGT complex.

Gel filtration

Gel filtration on Sephadex G–200 has been used to separate free toxin from receptor–toxin complex utilizing the difference in their molecular weights [84]. An improved method uses the cation exchange resin carboxymethylcellulose [85], thus utilizing the difference in charge between toxin and receptor for separation. [^{125}I]α-BGT is retained on the resin, while the complex is eluted in the resin minicolumns made in Pasteur pipettes. This method can be used for membrane-bound, solubilized as well as purified ACh receptors with complete recovery of reacted and unreacted species. A low blank is obtained consistently and the method is simple and rapid.

Ion flux

Since activation of ACh receptors was assumed to increase ion flux, attempts were made to recreate this mechanism *in vitro* by use of electric organ membranes, which form closed vesicles that could retain $^{22}Na^+$ as well as other cations [86]. The vesicle suspension in 0.6 M sucrose is mixed with $^{22}Na^+$ solution; after 10–20 h at 4°C the vesicles are loaded with $^{22}Na^+$. The efflux of $^{22}Na^+$ is monitored by diluting 0.1 ml 50-fold with salt solution and at given times the mixture is filtered rapidly on Millipore HAWP 0.25 μm filters, washed with cold buffer and the filters counted for radioactivity. Application of receptor agonists (carbamylcholine, ACh or suberyldicholine) caused an increase in the efflux of $^{22}Na^+$, $^{42}K^+$ and $^{45}Ca^{2+}$ but not $^{36}Cl^-$, while the presence

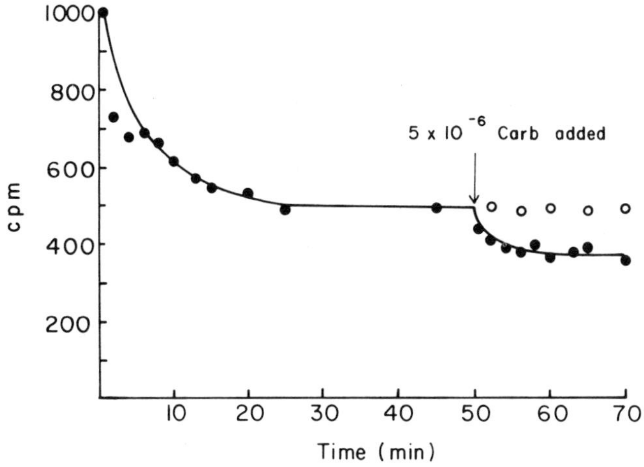

Fig. 5.6 Efflux of $^{22}Na^{+}$ from ACh receptor-enriched vesicles, formed from *T. ocellata* electric organs, as a function of time. After efflux reaches an almost steady state, addition of 5 μM carbamylcholine (●) causes an increase in $^{22}Na^{+}$ efflux, which is inhibited in the presence of 5 μM *d*-tubocurarine (○).

of antagonists (e.d. *d*-tubocurarine, flaxedil or cobrotoxin) inhibited this increase [86–88]. However, thc agonist-induced increase in ion flux from these vesicles was 100 000-fold lower than that in the endplate [24].

Hess and co-workers [89] modified and improved the efficiency of this technique (Fig. 5.6) by reducing the contribution of non-specific efflux; a technique that we have used. They were also able to separate the excitable from non-excitable vesicles (personal communication). Enrichment of the ACh receptor carrying membranes by the affinity partitioning method of Flanagan *et al.* [90] may further refine this method.

5.3.3 Purification of ACh receptors

Electric organ receptors

Most of the ACh receptors that have been purified are those of electric organs. The receptor molecules could not be removed from their membrane surroundings by repeated homogenization, sonication, or high salt concentrations, thus detergents (e.g. Triton X-100) were used [70, 91–93]. Use of gel filtration, electrophoresis, precipitation or isoelectrofocusing to purify ACh receptors succeeded only partially, but in the process the ACh receptor lost some of its ACh binding ability [94].

Affinity chromatography, however, which was introduced by Cuatrecasas [95], led to successful purification of ACh receptors. In this method, a ligand that is specific for the ACh receptor (e.g. α-cobrotoxin [96–99] or quaternary ammonium compounds [100–103] is covalently attached to Sepharose gel. When the detergent-solubilized electric organ membranes are incubated with this gel, either in a beaker or a column, the ACh receptor molecules plus a few others bind to the attached specific ligand. The non-specific proteins are removed by successive washing of the gel with high salt solutions, either by mixing with gel followed by filtration or by passing through the gel column. The ACh receptor is then desorbed by a high concentration of a specific ligand (e.g. carbamylcholine [96, 98, 100], hexamethonium [97], decamethonium [99], or NaCl [102]). After removal of this ligand by dialysis, the ACh receptor is in a relatively pure form, having specific binding of up to 10 [104, 105] or even 12 nmol/mg protein [100, 106]. By using the batch method rather than the column and by speeding up the steps, we are able to obtain high yields of purified ACh receptor 24 h after homogenization of the electric organ (Fig. 5.7). The speed plus the presence of chelating agent and anti-proteases during purification allow for the preservation of free SH groups and deter deterioration of the receptor molecule.

Another approach to the purification of ACh receptors has been that of extraction by chloroform-methanol followed by chromatography on Sephadex LH20 [107]. However, the proteolipid obtained is different from the protein purified by affinity chromatography. Antisera against the former do not cross-react with the latter and antisera against the latter do not cross-react with the former [108].

Muscle receptors

The nicotinic ACh receptor from rabbit skeletal muscle was also partially purified by affinity chromatography to a specific binding activity of 0.2 nmol $[^{125}I]$ α-BGT mg protein [109], which is 60-fold lower than the highest purity obtained for an electric organ ACh receptor. Their low concentration in the tissue makes purification and study of their molecular properties extremely difficult.

Brain receptors

The nicotinic ACh receptor of rat brain (detected by $[^{125}I]$ α-BGT binding) was solubilized with Triton X-100 and partially purified by affinity chromatography [110] from specific binding of 0.024 to 12 pmol/mg protein, almost one thousandth the purity achieved for the

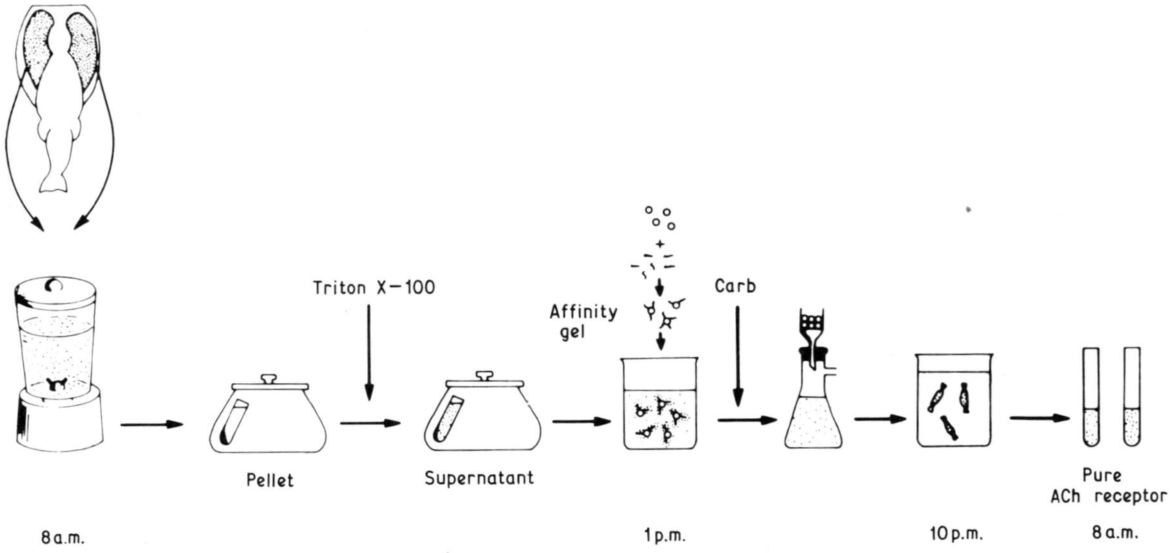

Fig. 5.7 Diagram of the protocol for purification of ACh receptors of *Torpedo* electric organ. After incubation of affinity gel with tissue extract, the mixture is filtered then the gel washed twice with buffer and twice with 1M NaCl and filtered to remove non-specifically adsorbed proteins before desorption of ACh receptor with Carb (1 M carbamylcholine).

electric organ receptor. It exhibited typical nicotinic pharmacology. The muscarinic receptor of the mammalian brain has not been purified, but successful solubilization with digitonin [111], while retaining its binding activity, should lead to its purification in the not too distant future.

Another ACh receptor that has been purified is the putative ACh receptor of housefly brain. These ACh receptors are solubilized simply by homogenization in water, and are purified by gel filtration and semi-preparative electrophoresis [112]. The fact that these receptors exhibit a nicotinic muscarinic pharmacology is not too surprising. ACh receptors having muscarinic-nicotinic pharmacology have been detected in mammalian brain stem [34] and thalamus [35]. Another putative invertebrate ACh receptor, from the squid *Loligo opalescens,* was also solubilized by homogenization [113]. The low dissociation constant for ACh binding to the isolated housefly putative ACh receptors (10 μM) and its mixed pharmacology agree with the data obtained by Sattelle from electrophysiologic studies (personal communication). The isolated housefly head proteins lost 90% of their binding during purification, but rabbit antisera produced against them did not cross-react with the nicotinic ACh receptor purified from *Torpedo* electric organ.

5.4 CHEMICAL NATURE OF ACh RECEPTORS

5.4.1 Composition of the electric organ ACh receptor

In 1952, Nachmansohn predicted that ACh receptors were membrane proteins [10], and twenty years later his prediction was verified. There is sufficient evidence that ACh receptors of mammalian brain, smooth or skeletal muscles as well as electric organs are proteins. The only ACh receptor that has been purified is that of electric organs, thus it is the only one whose composition has so far been studied.

The purified ACh receptor of electric organs is a glycoprotein, possibly containing 0.9–2 mol % hexosamines and other sugars, e.g. *d*-glucose [104], glucosamine [114], *N*-acetyl-*d*-glucosamine [104], *d*-mannose [104, 115] and *N*-acetyl-*d*-galactosamine [115]. Galactose was detected in one preparation [104] and not in another [115]. There is nothing extraordinary about the amino acid composition of the ACh receptor. It is made up of 20–26 mol % acidic amino acids and 11–12 mol % basic amino acids [97, 99, 104, 105, 115–117]. This overall high acidity is reflected in an isoelectric point of 4.5–4.8 [98, 99].

Study of the amino acid analyses of purified ACh receptors and comparison of the polarity index and average hydrophobicity, as well as the ratio of hydrophilic to hydrophobic amino acids, led to the conclusion that ACh receptors are 'integral' membrane proteins [118]. The receptor possibly penetrates substantially into the membrane core and is closely associated with lipids. Brisson *et al.* [119] used the choline ester of a spin-labeled fatty acid, 8-doxylpalmitoylcholine, which blocked carbamylcholine-induced depolarization of *Electrophorus* electroplax, slowed toxin binding and displaced [^{3}H] ACh bound to the membrane-bound ACh receptors. Electron spin resonance spectra measurements indicated that the spin label was completely immobilized as a result of its binding and was mobilized by cholinergic ligands, possibly by means of displacement of the spin label from the receptor into a fluid lipid phase.

The ACh receptor of electric organs has strong intrinsic fluorescence due to its content of 1.4–2.4 mol % *l*-tryptophan. When excited at 290 nm, the ACh receptor emits fluorescence with a maximum at 336 nm. Thus, as with most proteins [120] this tryptophan fluorescence is shifted to wavelengths shorter by 12 nm than that of free *l*-tryptophan.

The maximum number of free SH groups obtained in the receptor preparation is 20 nmol/mg protein (equal to 18% of the total cysteic acid residues) [105]. Preparations with less free SH groups are obtained if extra care is not taken during purification, and this reflects in lower specific binding.

Another component of the ACh receptor molecule is bound Ca^{2+}. We found that 4.7% of the molecular weight of the ACh receptor was due to bound Ca^{2+}, if purification was in Ca^{2+}-containing solutions, but was only 0.6% if EDTA-containing solutions were used [121]. This very high affinity binding of Ca^{2+} to the electric organ ACh receptor has recently been confirmed by Chang and Neumann [122]. The possible role of Ca^{2+} will be discussed in Section 5.5.2.

5.4.2 Molecular weight, shape and subunit structure

Every available purified ACh receptor preparation has detergent bound to the receptor at concentrations that vary from 11% to 50% of the total molecular weight of the receptor–detergent complex [104, 123, 124]. The presence of the detergent complicates the accurate determination of molecular weight by adding weight and buoyancy to the molecule. Molecular weight estimates obtained by sucrose density gradient centrifugation in H_2O and D_2O are 360 000 and 320 000 for the Triton X-100- and deoxycholate- solubilized electric organ ACh receptor preparations respectively [124]. The same methof gave a molecular weight value of 357 000 for the Triton-solubilized rat brain nicotinic ACh receptor [110]. Another method used is sodium dodecylsulfate (SDS) disc gel electrophoresis of the α-BGT-labeled ACh receptor but, since this detergent dissociates α-BGT from the receptor, prior fixation of the receptor with glutaraldehyde or suberimidate is required. Values of 260 000 [99] and 275 000 [125] were obtained, respectively. Membrane osmometry gave a molecular weight of 270 000 [126]. Sedimentation equilibrium studies by Edelstein *et al.* [123] for ACh receptor preparations that had relatively little bound Triton showed two molecular species, a 330 000 and a 660 000, but when 0.1% Triton was added only the 330 000 species remained (see Fig. 5.8), suggesting that 0.1% Triton either prevents aggregation of two receptor molecules or splits one molecule into two. By sedimentation in sucrose gradients of α-BGT-labeled crude and pure ACh receptor of *T. californica,* O'Brien and collaborators [127] detected several molecular weight species; the major ones being 190 000 and 330 000 and the minor ones 80 000 and 500 000. These data point to the heterogenity of molecular weights of oligomers and ease of aggregation of ACh receptors.

Sodium dodecylsulfate electrophoresis is used to separate the subunits of the ACh receptor after its denaturation. Except for a single report, which suggests that the ACh receptor molecule is made up of only one kind of subunit [97], several others find two major subunits plus one, two or more minor ones [79, 103, 104, 115, 125, 128, also Fig. 5.9]. The variability observed in the number and size of the subunits may be a result of differences in the purity of the receptor preparation, content of detergent, denaturation treatment and even the method of SDS disc gel electrophoresis used [115]. Proteolysis may also occur during purification, and this results in smaller subunits [129].

There is general agreement that the smallest molecular weight subunit ($\simeq$ 40 000), which represents the largest amount of protein, carries binding sites for antagonists and agonists. This is based on the finding that the active site-directed reagent, [^{3}H]*p*-(*N*-maleimido)-2-benzyltrimethylammonium, [106, 115] photoaffinity label, 4-azido-2-nitrobenzyltrimethylammonium fluoroborate [130], covalently bound to it and blockade of these bindings by cobrotoxin [115].

The second smallest molecular weight subunit does not bind the maleimido compound nor cobrotoxin [106, 115]. The larger size

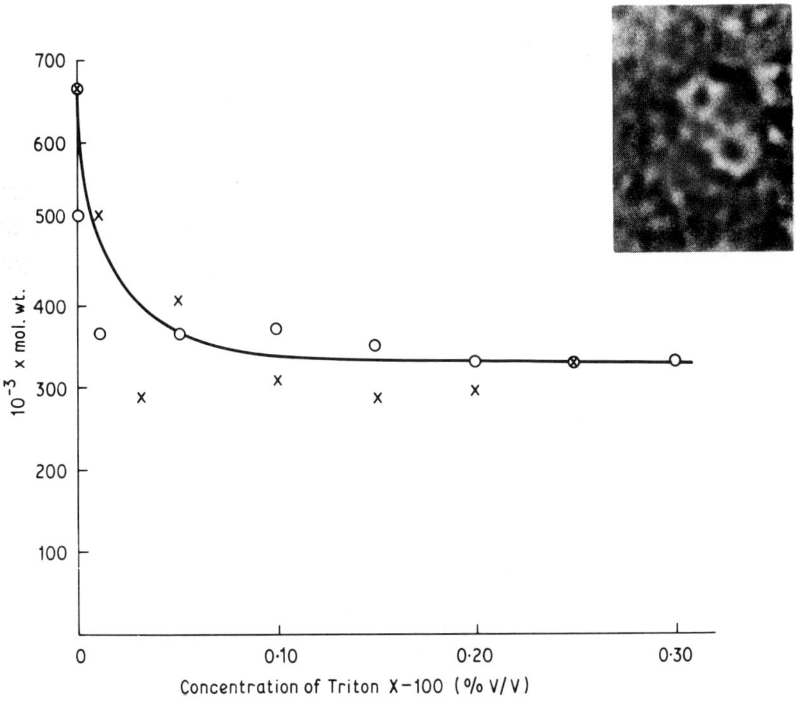

Fig. 5.8 Molecular weight of purified receptors from *T. californica* (○) and *T. marmorata* (x) determined by sedimentation equilibrium in the presence of increasing concentrations of Triton X-100 (from Edelstein *et al.* [123]). *Inset* is electron micrograph of negatively stained purified ACh receptor of *T. californica* (1093 000 x magnification) (from Eldefrawi *et al.* [131]).

molecular weight subunits may or may not be a product of incomplete hydrolysis.

If the receptor is treated mildly with trypsin the size of all the subunits becomes less than 40 000 [131], and if the treatment is stronger and with chymotrypsin and trypsin the largest subunit becomes 8000 in molecular weight [132].

If it is assumed that the molecular weight of the electric organ ACh receptor is 320 000, and the monomer (subunit(s) that carries one ACh-binding site) is 80 000 (calculated by dividing the mg protein by the moles of specific binding sites), then there would be four ACh-binding sites per receptor molecule. If the molecular weight of the receptor in the membrane is 660 000, then this value would double to eight ACh-binding sites. Although still controversial, it is believed by many that the electric organ ACh receptor carries an equal number of sites for ACh as

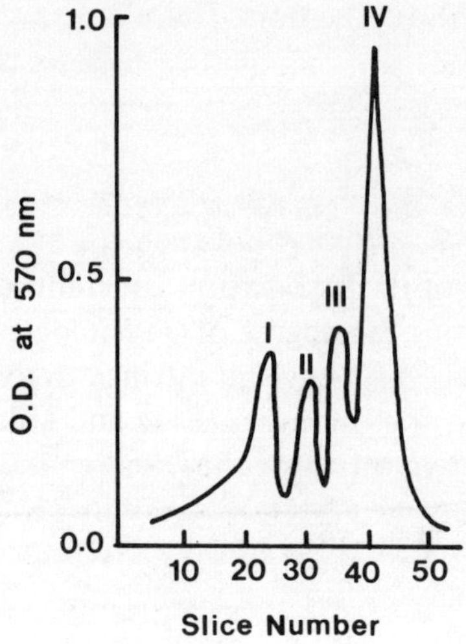

Fig. 5.9 SDS disc gel electrophoresis of the ACh receptor purified from *T. ocellata* electric organ. The apparent molecular weight of the subunits, as calculated by comparison with calibrated proteins are I, 81 000; II, 61 000; III, 50 000 and IV, 40 000.

electric organ ACh receptor carries an equal number of sites for ACh as for α-BGT [105, 115].

Electron micrographs of the negatively stained membrane-bound or purified electric organ ACh receptor show it as a doughnut with an electron-dense core [131, 133, 134, also Fig. 5.8]. However, similarly shaped proteins are found in many other membranes where no ACh receptors are known to exist, such as luminal membrane of the urinary bladder and gap junctions of liver [135] and goldfish brain [136]. It is plausible that this array is associated with structures regulating ion fluxes in special membranes, and in the case of cholinergic postsynaptic membranes the molecules are modified by the addition of ACh-binding sites plus other characteristics that make it the ACh receptor. The ACh receptor of electric organs appears to be organized in a lattice in the plane of the membrane as observed by X-ray diffraction [137].

5.4.3 Binding sites

According to classical pharmacology, drug–receptor interactions have been explained by several hypotheses. There is the 'rate theory' of Paton [138] which postulates that it is the rate at which the drug–receptor complexes are formed and broken that determined the type as well as the time-course of drug action. There is also the 'occupation theory', which postulates that drug binding follows the Langmuir isotherm and that the response is directly proportional to the percentage of receptors occupied [139–141]. Since different ligands induced different degrees of the same response, Stephenson [139], Ariens [140] and Furchgott [141] differentiated between the affinity of the drug for the receptor (reciprocal of its dissociation constant) and its 'efficacy' or 'intrinsic activity', which is the degree of physiological response the drug is able to induce. Thus agonists have an intrinsic activity, while antagonists do not. However, the finding of Katz and Miledi that the degree of response to a cholinergic activator is related to the length of time the channel remained opened in response to the interaction of the drug with the ACh receptor (see Section 5.2.1), offers a molecular explanation for intrinsic activity.

The rate and occupation theories assume that agonists and antagonists bind to the same site on the receptor molecule. However, biochemical studies on membrane-bound solubilized or purified ACh receptors are producing data that may modify our thinking. For example, drug-receptor interactions, like enzymes, may deviate from following Langmuir isotherms and may exhibit co-operative interactions (see Section 5.5.1). In addition, equilibrium and kinetic studies suggest that the ACh receptor carries sites for binding agonists, which are different from the ones that bind antagonists (Fig. 5.10). Binding of [^{3}H]ACh is blocked competitively by the presence of the agonists nicotine or carbamylcholine and blocked non-competitively by the presence of the antagonist *d*-tubocurarine [142]. Kinetics of the binding of [^{125}I]α-BGT to membrane-bound ACh receptors also indicate that carbamylcholine occupies separate binding sites on the receptor molecule from the ones binding *d*-tubocurarine and α-BGT and only half of the sites interact with each other [143, 144].

The sites that bind agonists and those that bind antagonists can be selectively inhibited. We found that methylmercury (at 10 μM) blocked [^{3}H]ACh binding to the electric organ ACh receptor essentially irreversibly [145, 146], without affecting α-BGT binding (Eldefrawi, unpublished). In addition, propidium, a fluorescent organic compound, was found to bind to the purified *Torpedo* receptor with a 15-fold fluorescent enhancement and K_D of 1.4 μM. *d*-Tubocurarine and hexamethonium, but not agonists, blocked this binding competitively (Sharon, Hess and Eldefrawi, unpublished). Raftery and co-workers [147] were able to modify one anionic binding site on *T. californica* ACh receptors (in membranes, solubilized or purified) with trimethyloxonium tetrafluoroborate and affect binding of agonists and small antagonists such

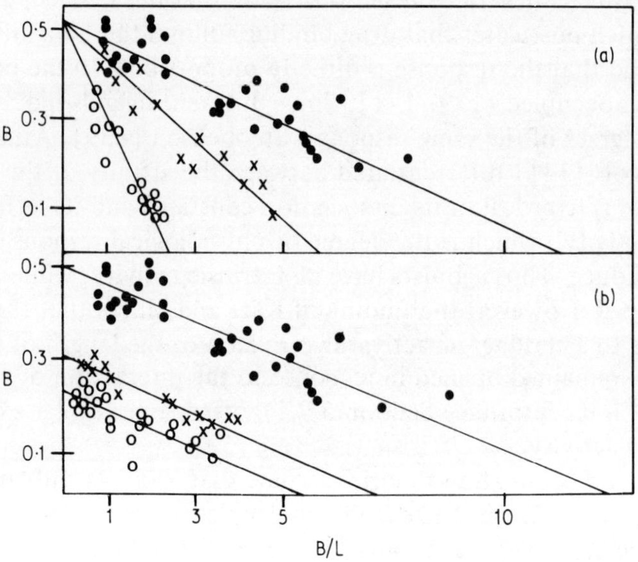

Fig. 5.10 Scatchard plot of the binding of [^{3}H] ACh to the membrane-bound ACh receptor of *T. marmorata,* determined by equilibrium dialysis. (a) Binding of [^{3}H] ACh alone (●) and in the presence of 1 μM carbamylcholine (x) or 10 μM nicotine (○). (b) Binding of [^{3}H] ACh alone (●) and in the presence of 10 μM gallamine (x) or 1 μM *d*-tubocurarine (○). *B*, amount of [^{3}H] ACh bound in nmole/g tissue; *L*, concentration of free [^{3}H] ACh. Common intercept at the ordinate suggests competitive binding.

as *d*-tubocurarine, but not of α-BGT. They suggested that the site most likely modified might be a carboxyl group adjacent to the ACh binding site; a site that was also suggested as being involved in binding the second trimethylammonium group in decamethonium. The proposal that electric organ ACh receptors bind agonists and antagonists at different site is not widely accepted. In fact, binding data have also been interpreted by means of an allosteric model for co-operativity extended to half-of-site activity [148].

There are other different kinds of sites, which are detected in the electric organ ACh receptor molecule. There are the antigenic determinant sites, most of which are not at the ACh- or α-BGT-binding sites [149–152]. There are also $^{65}Zn^{2+}$-binding sites (56 nmol/mg protein as compared to 10 nmol ACh-binding sites/mg protein, a third of which may be free SH groups) and Ca^{2+}-binding sites (a minimum of 150 nmol/mg protein) [121]. There is also the carbohydrate portion of the receptor molecule which binds plant lectins, as shown for the ACh receptor of electric organs [104, 114, 115] and that of skeletal muscle [153]; and this binding inhibits α-BGT binding.

The muscarinic smooth muscle ACh-receptor may also have different binding sites. Studies on the inhibition of ACh-induced contractions of the guinea-pig ileum led Triggle and collaborators [154, 155] to suggest that aryldiazonium salts and benzilylcholine bound to the agonist recognition site plus another allosteric site which decreased the affinity of the receptor for agonists.

5.5 MOLECULAR MECHANISM OF ACh RECEPTOR ACTION

5.5.1 Co-operative interactions and conformational changes

Electrophysiologic experiments detected positive co-operativity in the effects produced by agonist interactions with the ACh receptors of skeletal muscle and electric organs [156, 157]. In its simplest form, positive co-operativity is the phenomenon where binding of a molecule to a protein enhances the binding of a second molecule to the same protein. The co-operativity observed could be due to binding of agonists to the ACh receptor or could occur in subsequent steps (e.g. activation of the receptor channel) that end with the change in membrane potential. However, binding of [^{3}H]ACh to the electric organ ACh receptor indeed exhibits positive co-operativity, whether the receptor is in the membrane [105, 158-160], solubilized [148, 158] or purified [158]. The Hill coefficient obtained of 1.6–2.0 suggests that there are at least two agonist binding sites per ACh receptor molecule and quite possibly 4 or 6 [148], which agrees with the specific binding data.

Co-operative binding of ligands to a protein which has more than one binding site for activators (oligomeric protein) has been explained by one of two models. The two-state (or allosteric) model of Monod, Wyman and Changeux [161] where the protein oscillates spontaneously between two conformations (i.e. three-dimensional structure) with differing affinities for ligands. One state is favored when the protein is free or binds an antagonist and the equilibrium between the two states shifts towards the other when the protein binds an agonist. This model has been applied to the evaluation of electrophysiologic [162] and biochemical [148] data on ACh receptors. The induced-fit model of Koshland and co-workers [163] proposes that binding of a ligand to one subunit of the protein induces changes in the conformation of the protein which make it either easier or more dificult for a second ligand to bind to the other subunit. This model was utilized to explain kinetics of the binding of α-BGT to the electric organ receptor [144].

Co-operativity in binding may be indicative of a conformational changes comes from *Naja nigricollis* α-[^{3}H] toxin binding [164, 165] and studies of Hess *et al.* [144] with [^{125}I]α-BGT binding to ACh receptors of electroplax membrane preparations from *E. electricus.* The reaction between toxin and receptor was initially fast followed by a slower one. Since the fraction of the reaction in the fast phase increased with increasing initial toxin concentrations, while the fraction going slowly decreased correspondingly, and bot phases were affected by cholinergic ligands, it was proposed that at least two different toxin-binding sites existed on the receptor, and toxin binding induced the conversion of one site into the other.

In studies by Sarvey *et al.* [166] the essentially irreversible effect of α-BGT on mammalian skeletal muscle was converted into a reversible effect by an activating or a desensitizing dosage of the agonist carbamylcholine. This effect is best explained by a change in receptor conformation which is induced by the agonist resulting in the reversible interaction. It is equally possible that there are different sites for agonists and tagonists on the ACh receptor, which are directly coupled, so that binding of agonist to one site allosterically affects the affinity of the inhibitor-binding sites and *vice versa.*

The present assumption is that binding of ACh to its receptor induces a change in the conformation of the receptor, which in turn causes a channel to open and ions to flow through it. Not only does the ACh receptor change its conformation upon binding of an agonist, but it appears also to change into a third conformation when in a desensitized state (see Section 5.5.6). An extremely useful tool for detection of such changes is the study of fluorescence, either the native fluorescence of the ACh receptor in electric organ membranes utilized by Barrantes and co-workers [167, 168], or the fluorescence of probes utilized by Changeux and co-workers such as the local anesthetic quinacrine [169] or 1- (5-dimethylaminonaphthalene-1-sulfonamide) ethane-2-trimethylammonium iodide (DNS-chol) [170]. Different changes in fluorescence of the receptor or receptor–probe complex occur upon binding of an agonist to the receptor or upon desensitization of the receptor molecule.

Further evidence in support of conformational changes occurring in the receptor molecule is the apparent decrease in ACh affinity that is observed after pre-exposure of the membrane-bound electric organ receptor to ACh (for details see Section 5.5.6). The release of bound Ca^{2+} by the purified ACh receptor upon binding of ACh or agonists

[106, 122, 132, 171] and the re-uptake of Ca^{2+} by subsequent addition of α-BGT [122] also led to the suggestion that ACh and α-BGT caused different conformational shifts in the ACh receptor [122].

5.5.2 Role of calcium in ACh receptor function

Calcium has a regulatory function in both chemically and electrically excitable systems [172–175]. An essential role for it has been established in the coupling of excitation to smooth muscle contraction [173, 174]. ACh sensitivity is affected by Ca^{2+} removal or addition of lanthanum [175], leading to the suggestion [173] that the agonist–receptor complex activated Ca^{2+} translocation machinery that supplied the contractile element of the muscle with Ca^{2+}. Chang and Triggle's studies [174] on guinea-pig ileum led them to add that intrinsic activity of an agonist was a direct reflection of its ability in the agonist–receptor complex to activate the membrane-bound Ca^{2+} translocation system. The source of mobilized Ca^{2+} was uncertain, but it was suggested that a labile membrane-bound source of Ca^{2+} (in equilibrium with extracellular Ca^{2+}) was utilized as well as extracellular Ca^{2+}. Whether Ca^{2+} played a role in the agonist–receptor interaction or in coupling to muscle contraction was not known from the electrophysiologic data.

With the availability of a purified electric organ receptor, evidence started accumulating in support of a role for Ca^{2+} in the reaction of ACh receptor to binding of agonist. The ACh receptor was found to have a very high affinity for Ca^{2+} [121, 122; also see Section 5.4.1]. In addition to this high affinity Ca^{2+} binding, Martinez-Carrion and Raftery [176] reported that binding of Ca^{2+} inhibited receptor binding of a fluorescent ligand with an inhibition constant (K_i) in the mM range. We obtained similar results by inhibition of [^{3}H] ACh binding [121].

The fluorescent lanthanide, terbium (Tb^{3+}), bound to the purified electric organ ACh receptor caused a fluorescent enhancement of 10^4. Because about 60% of the Tb^{3+}-binding sites also bound Ca^{2+}, we could determine the effect of agonists and antagonists on these Ca^{2+} binding sites. We found that binding of agonists caused release of bound Tb^{3+} from the Ca^{2+}-binding sites (3–6 Tb^{3+} ions released per ACh receptor site), while binding of inhibitors did not (Fig. 5.11). Chang and Neumann [122] reported similar data using murexide as a sensitive Ca^{2+} indicator, but found that the binding of ACh to the electric organ ACh receptor molecule caused it to release 2–3 Ca^{2+} ions per ACh molecule bound, and subsequent addition of α-BGT caused a re-uptake of Ca^{2+}. Displacement

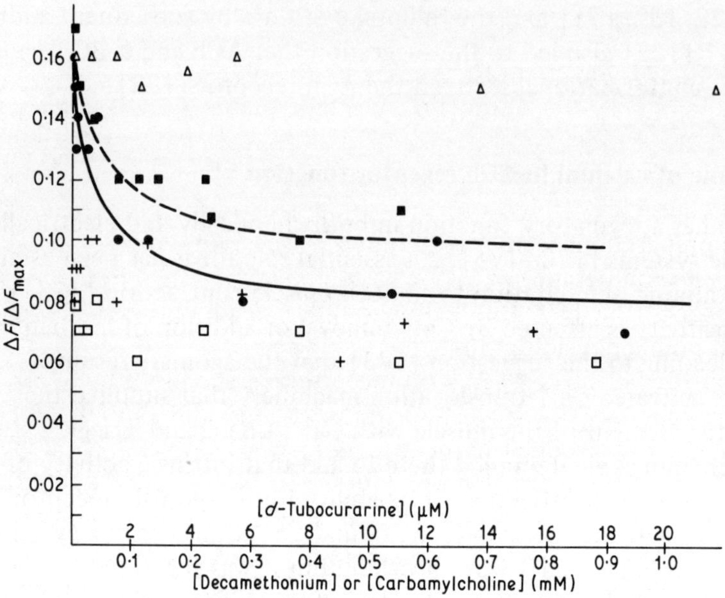

Fig. 5.11 Amount of Tb^{3+} bound ($\Delta F/\Delta F_{max}$) to ACh receptor, purified from *T. ocellata,* in presence of increasing concentrations of drug *d*-tubocurarine (△), decamethonium (●), carbamylcholine (■), decamethonium in presence of 8 mM $CaCl_2$ (+) or carbamylcholine in presence of 8 mM $CaCl_2$ (□). Bound Tb^{3+} is displaced from the receptor by decamethonium or carbamylcholine but not *d*-tubocurarine. When the sites that bind Tb^{3+} or Ca^{2+} are occupied by Ca^{2+} (i.e. in presence of 8 mM $CaCl_2$), little or no Tb^{3+} is displaced by the receptor agonists, thus suggesting that the displacement of Tb^{3+} is from the Ca^{2+}-binding sites (from Rübsamen *et al.* [132].

of Ca^{2+} from the ACh receptor by ACh or agonists may be directly responsible for receptor-mediated changes in permeability of neural membranes to inorganic ions.

The Ca^{2+} and Tb^{3+}-binding sites were located on the 40 000 subunit of the *Torpedo* ACh receptor. This binding was preserved even after degradation of the receptor by trypsin and chymotrypsin to peptides of 8000 daltons or less. Thus, it was suggested that the Ca^{2+}-binding sites were determined by structural features of the peptide chain rather than by the three-dimensional arrangement of the intact receptor [121]. Whether an exotic amino acid, such as γ-carboxyglutamic acid, is responsible for Ca^{2+} binding to the ACh receptor, as was found in prothrombin [177, 178], is still unknown.

5.5.3 Reconstitution of the ACh receptor

There is no *a priori* reason to expect that the molecule that binds ACh (i.e. the ACh receptor) is the same molecule that is directly responsible for ion translocation. Nevertheless it was assumed by many, including ourselves, that it may, and attempts were made to reconstitute the purified nicotinic ACh receptors in synthetic lipid membranes. Partially purified preparations were incorporated in black lipid bilayer and increases in conductance of Na^+, K^+ or Cl^- were obtained in the presence of activators [109, 179, 180]. However, in no case did the excitability obtained mimic physiologic events.

The purified electric organ ACh receptor was successfully incorporated in lipid vesicles by Michaelson and Raftery [181] and an increase in $^{22}Na^+$ efflux was obtained in the presence of carbamylcholine, but not if α-BGT was also present. However, this increase was only 10-fold, cation selectivity was not demonstrated, and only 10% of the ACh receptor preparations yielded chemically excitable vesicles. Using bilayers of oxidized cholesterol, Shamoo and Eldefrawi [182] formed the membrane in presence of purified electric organ ACh receptors and ACh or carbamylcholine. We detected increased monovalent cation conductances that varied from 10 to 100-fold, but this occurred consistently only when the receptor had been treated mildly with trypsin. The observed response was also very slow, taking minutes instead of a second or less to develop after the addition of the agonist. Once conductance was increased, it could not be stopped by subsequent addition of curare or α-BGT; only if the antagonist was present with the receptor before addition of agonist was there no selective increase in conductance. Two other attempts to reconstitute this ACh receptor preparation and another preparation from *Torpedo californica* [183, 184] in phospholipid vesicles failed to produce agonist-induced cation selective flux.

It is possible that these purified ACh receptor preparations lack a component that is necessary for proper reconstitution, or maybe the correct conditions for their reconstitution have not been met.

5.5.4 The ACh receptor and its ion conductance modulator

We may visualize the ACh receptor as a protein which penetrates through a bilayer membrane of phospholipids and cholesterol. This protein plays two roles, structural by contributing to the integrity of the membrane and functional by translating the binding of ACh to its recognition site into a permeability change in the membrane. It is probably closely associated in the membrane with an entity, which has been called by Albuquerque *et al.* [185] the 'ionic conductance modulator' (ICM). Its molecular nature and its morphological relationship to the ACh recognition site are as yet not fully known. It could be part of the ACh receptor molecule, or a separate molecule that is closely associated with the receptor through non-covalent bonds as recent data suggest. Thus, the ICM may be the ionic channel (a component that transverses membrane through which ions pass) or ionophore (a small molecule which shuttles across the membrane to transport ions) as well as any of the structures that link these entities to the ACh receptor.

Histrionicotoxin (HTX), an alkaloid isolated from the skin extracts of the Colombian frog, *Dendrobates histrionicus,* and its completely reduced analog, perhydrohistrionicotoxin (H_{12}-HTX) were utilized in electrophysiologic studies of mammalian and amphibian nerve–muscle preparations [186–190]. They reduced the amplitude of endplate current and shortened half decay time. Their main target was suggested to be the ICM of the ACh receptor. Kato and Changeux [191] found that HTX and dihydro-HTX blocked depolarization of the electroplax by carbamylcholine without blocking binding of [^{3}H] ACh to the *Torpedo* membrane-bound ACh receptor. In fact, at concentrations below 1 μM, HTX increased binding of [^{3}H] ACh. Since some local anesthetics had similar effects on binding of the membrane-bound (but not the solubilized) ACh receptor [192], it was suggested that both HTX and these local anesthetics bound to an allosteric site on the ACh receptor or to a closely associated molecule [191].

When binding of [^{3}H] H_{12}-HTX to *Torpedo* electric organ membranes was studied, it bound with a high affinity (K_D = 0.4 μM) to twice the number of sites that bound [^{3}H] ACh or [^{125}I] α-BGT and the binding was blocked by receptor agonists as well as some local anesthetics, but not by receptor antagonists. No such binding to the purified ACh receptor was detected. Upon solubilization of the ACh receptor, agonists could no longer block [^{3}H] H_{12}-HTX binding, but local anesthetics and HTX derivatives still did, though to a lesser degree [193]. We suggested that this high affinity binding was to the ICM of the ACh receptor, which was separated from the receptor by solubilization with Triton X-100, so that HTX binding was affected by receptor agonists only when receptor and ICM molecules were closely associated in the membrane. The ACh receptor and the [^{3}H] H_{12}-HTX binding protein molecules (putative ICM) were totally separated by immunoprecipitation of the receptor or its adsorption by affinity gel, and partially separated by gel filtration [193].

A reasonable, but unsupported, suggestion is that the two molecules may be coupled by Ca^{2+}, particularly in view of the high affinity the receptor has for Ca^{2+}. Since binding of agonists to the receptor causes it to release Ca^{2+}, it may be that Ca^{2+} release results in a change in the dipole moment of the ICM molecule leading to opening of the channel. Since $[^3H]H_{12}$-HTX does not bind with a high affinity to the purified ACh receptor it probably lacks part or all of the ICM. If so, it would account for the failure to reconstitute the purified ACh receptor in a system that mimics physiologic conditions. Partial successes in this respect may be due to the presence of ICM as impurities in the purified ACh receptor preparations. We may predict that reconstitution of ACh receptor and the ICM after its purification should provide us with the desired system.

5.5.5 Activation of nucleotide cyclase, protein phosphorylation and turnover of phospholipids

The discovery by Greengard and Ferrendelli and their collaborators that ACh and/or muscarinic drugs (but not nicotinic ones) caused accumulation of guanosine 3′: 5′-cyclic monophosphate (cGMP) in heart [194, 195], brain [195, 196], superior cervical ganglion [197, 198], smooth muscle [199, 200] and neuroblastoma [201], and its inhibition by atropine, suggested that cGMP might be acting as a second messenger and mediate the action of muscarinic transmission. cGMP activates protein kinases, and cGMP-dependent protein phosphorylation has been detected in different types of smooth muscle ([200, 206], also see Greengard [202]). Since agonist activation of cGMP was dependent upon the presence of extracellular Ca^{2+}, it was suggested that Ca^{2+} flux may participate in the control of intracellular cGMP [199]; also that cGMP-phosphorylated protein may activate Ca^{2+} permeability, which supplies the contractile machinery of the smooth muscle with Ca^{2+}, as has been suggested by Chang and Triggle [174].

There appears to be a link between ACh receptor (mainly muscarinic) stimulation and the agonist-stimulated turnover of a phospholipid. ACh or carbamylcholine have been shown to increase incorporation of $^{32}P_i$ into certain phospholipids in slices of cerebral cortex and in sympathetic ganglia [203–205], and so has electrical stimulation [206]. Atropine blocked it [207] and it occurred preferentially in synaptosomes [204]. This incorporation was also detected in electric organ [208] and ileum smooth muscle [209]. Thus, the phenomenon was interpreted to be mainly postsynaptic, though it was also found to occur pre-synaptically in vesicles [210]. There is disagreement as to the kind of phospholipid into which $^{32}P_i$ incorporation is stimulated by agonists. Evidence has been presented in favor of the following phospholipids: phosphatidylinositol [203, 208, 209], phosphatidic acid [203, 210] and phosphatidylcholine [208]. The rate of disappearance of [^{32}P]phosphatidic acid in synaptosomes of guinea-pig cerebral cortex was found by Schacht and Agranoff to be stimulated by 80% with the addition of ACh [203].

The close association between phospholipids and ACh receptors in the membrane was demonstrated by finding that binding of agonists to the electric organ receptors was partially inhibited by treatment with phospholipase C [69], and so was [^{3}H]QNB binding to the muscarinic brain ACh receptors (Aronstam and Abood, personal communication). Also, addition of phosphatidylserine was found to enhance specific binding by mammalian muscarinic brain ACh receptors by up to 40%, while phosphatidic acid and phosphatidylinositol were less effective, and natural lipids were without effect. [^{3}H]QNB binding was inhibited by addition of unsaturated fatty acids or phospholipids.

5.5.6 Desensitization

Desensitization is the phenomenon where there is loss of response to a previously effective dose of an agonist. It results from exposure to a high dosage of the agonist or prolonged exposure to low depolarizing dosage. It occurs in response to externally applied ACh or drugs or by excessive nerve stimulation [211, 212]. It was detected in skeletal muscle [211–215], smooth muscle [216] and in electric organs [217, 218]. The rate and extent of pharmacologic desensitization varies with the structure and concentration of the agonist used [211] and is enhanced by Ca^{2+} and other cations [219, 220] as well as by high temperature [221]. Some antagonists, having a higher affinity for the desensitized state, may also cause desensitization [222]. Efflux of $^{22}Na^+$ from electric organ microsacs is also decreased by preincubation with agonists [223].

Katz and Thesleff [211] proposed that desensitization was due to a slow reversible change in the conformation of the ACh receptor–agonist complex to an inactive form. Others have added that the desensitized ACh receptor has a conformation that is different from the resting form of the receptor; thus the receptor molecule exists in one of three conformations at any one time [169]. The affinity of the receptor for agonists would be highest in the desensitized state, lowest in the resting state and intermediate in the activated state. Antagonist binding would stabilize the first two states but not the last one [169].

The high concentration of agonist required to produce desensitization would suggest that it occurs through binding to a low affinity site. Reduction of binding of ACh to the electric organ receptor at high ACh concentrations was suggested to be related to desensitization [224]. Also, others have found that prolonged exposure to cholinergic ligands primarily resulted in a decrease of the intrinsic ability of the receptor protein to bind α-neurotoxins [225, 226].

An alternative view now gaining acceptance is that presented by Changeux. It was first observed that local anesthetics, or Ca^{2+}, which were shown to enhance desensitization *in vivo* [227] and *in vitro* [228], stabilized the membrane-bound ACh receptor in a state that bound ACh with a high affinity [229]. It was suggested that the desensitized ACh receptor had a high affinity for agonists. Supporting evidence was provided by studies of Weber *et al.* [224] on binding of *Naja nigricollis* α-[^{3}H] toxin to membrane-bound *Torpedo* ACh receptors. They discovered that pre-exposure to agonists caused a slow and reversible decrease in the initial rate of cobra toxin binding when agonist was present with the toxin but not when the agonist was absent. The antagonists flaxedil and *d*-tubocurarine did not cause such a change, but hexamethonium and decamethionium did. It was proposed that, in the membrane at rest, the receptor was present in a state with low affinity for agonists, and that agonists stabilized the receptor in a conformation having a high affinity for agonists, a conformation that corresponded to desensitization of the system.

The initial reduction in toxin binding as a result of preincubation with carbamylcholine occurred only when carbamylcholine was present with the toxin, leading to the suggestion that the affinity of the ACh receptor was increased for agonist and not decreased for toxin [224]. However, Sarvey *et al.* [166] found that exposure of the rat skeletal muscles to agonists caused the binding of α-BGT to the ACh receptor (detected by studying endplate depolarization) to change from essentially irreversible to reversible binding. Thus, part of the reduction in toxin binding observed by Weber *et al.* may be due to a change in the characteristics of toxin binding in the presence of agonist. These observations do indicate that changes in the conformation of the ACh receptor are induced by agonists. Total understanding of the molecular mechanism of desensitization of the ACh receptor awaits further investigations.

No changes in agonist binding properties of brain muscarinic ACh receptor preparations have been observed [230], but a decrease in agonist binding affinity of muscarinic receptors in the guinea-pig ileum smooth muscle was observed upon exposure to a high concentration of carbamylcholine [231]. A reduction in binding of [^{3}H] ACh to *Torpedo* ACh receptors was also detected upon exposure to high concentrations of ACh [232].

5.6 ACh RECEPTOR DEVELOPMENT AND TURNOVER

5.6.1 Development

The availability of cell cultures and the ease of controlling their environment has allowed studies of the development of ACh receptor in muscle cells in the absence and presence of neural influence. These studies are also feasible because of the availability of an almost irreversibly binding ligand such as α-BGT that is very specific for nicotinic ACh receptors. The ACh receptor is an interesting molecule for study of development because it undergoes large changes in concentration during muscle development in the embryo as well as after elimination of its nerve supply (i.e. denervation). Embryonic skeletal muscle fibers are sensitive to ACh along their entire surface, because the distribution of ACh receptors is equally diffused [233]. On the other hand, in innervated muscles, ACh receptors are mostly confined to the synaptic areas [234, 235]. Thus, an important question in developmental neurobiology has been whether or not synthesis and distribution of ACh receptors in muscle cells requires an input from neurons. This was definitively answered by a 'not' by several scientists, starting with the work of Nirenberg and co-workers [236], who discovered ACh receptors (identified by their specific binding of [^{125}I] α-BGT) in skeletal muscle cell cultures in the absence of neurons. Though the ACh receptors may not require neuronal input to be synthesized and distributed, they are influenced by it, as discussed below.

Developing skeletal muscle cells in culture appear at first as dividing mononucleated cells (myoblasts) and then fuse and form non-dividing large multinucleated cells (myotubes). The latter develop a striated contractile apparatus [237], show contraction and electrical activity in response to ACh [238] and respond to ACh by a fast depolarizing response [239].

As the myotube develops, ACh receptors are first evenly distributed

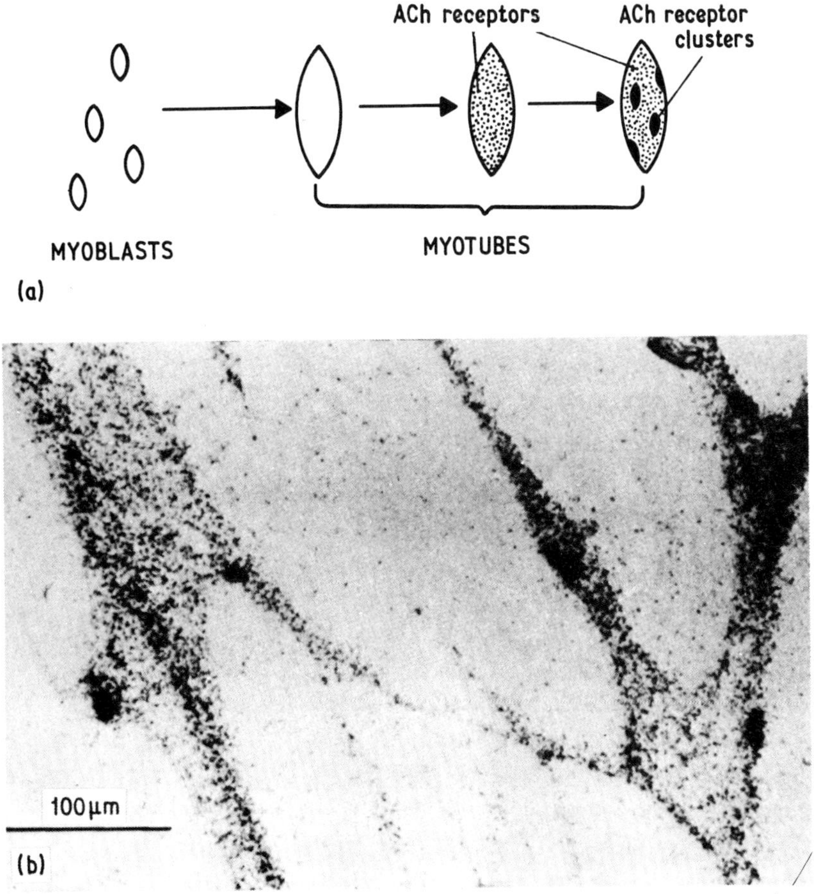

Fig. 5.12 (a) Diagrammatic representation of the development of ACh receptors in skeletal muscle cells in culture. (b) Autoradiograph of muscle cells labeled with [^{125}I] diiodo-α-BGT. Dark spots represent the labeled ACh receptors at low concentrations and in clusters (from Sytkowski *et al.* [240]).

over the entire surface, but later concentrate in clusters of $\simeq$ 9000 sites μm^{-2}, while the density of receptor sites in the surrounding areas is $\simeq$ 900 μm^{-2} [240] (Fig. 5.12). This compares to a density of 20 000–40 000 μm^{-2} for the ACh receptors in an endplate [17, 18]. The biochemical evidence for the cluster distribution of ACh receptors in these skeletal muscle cell cultures agrees with the electrophysiological data, where there are areas along the myotubes called 'hot spots', which have elevated ACh sensitivity ($\simeq$ 5 x higher than surrounding areas [241]). Similar hot spots had been observed in denervated skeletal muscles [242, 243]. Ultrastructural study of the membranes proved that the increased ACh sensitivity was not due to increased folding of the membranes, thus had to be due to increased ACh receptors per unit membrane [244]. It has recently been reported that there is another stage in ACh receptor development following cluster formation where in well-differentiated, cross-striated myotubes, there is a marked reduction in ACh receptors in the absence of neuronal influence [245]. ACh receptors were also found on neuron cells in culture [246, 247]. However, in contrast to the above cells the distribution of ACh receptors in cultured sympathetic ganglion neurons appeared to be uniform [248].

Since the source of muscle for primary culture is usually from 10–11 day old chick embryos, motoneuron axons are already in contact with the developing muscle, therefore, aneural hindlimb buds of chick embryo were used for cell cultures in one study to assure the absence of nerve influence [249]. Hot spots were still found, thus leading to the conclusion that their development was not dependent on any prior contact with neurons. ACh receptors were also detected in chick embryo retina before synapses appeared [250]. A problem that is still unresolved is whether the hot spot areas play any role in attracting the nerve axon, or whether a new spot forms at the synapse after its formation.

The development of muscarinic receptors has recently been followed in avian brain [251] and heart [252] by utilization of the specific binding of $[^3H]$QNB. Concentrations of receptor sites have been determined, but their distribution on cell surface awaits further study.

5.6.2 Effect of denervation on ACh receptors

In innervated mammalian muscle, ACh receptors are concentrated in the synaptic area. In a fast twitch muscle (e.g. extensor digitorum longus) ACh sensitivity (which reflects numbers of ACh receptors per unit area) is restricted totally to the endplate area. However, in a slow twitch muscle (e.g. soleus) sensitivity to ACh is highest at the endplate, but can also be detected outside the endplate along the entire muscle membrane [243, 253, 254]. Cutting off a piece from the innervating neuron (i.e. denervation) causes in 24 h simultaneous increase in ACh sensitivity along the membrane of both muscles that does not develop centrifugally from the endplate region. In the extensor, ACh sensitivity appears first close to the endplate and at the muscle tendon region (Fig. 5.13), while in the soleus, there is a simultaneous increase of ACh sensitivity along the entire

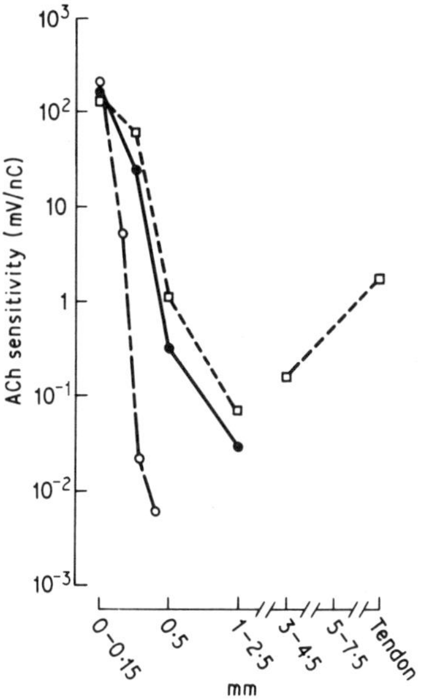

Fig. 5.13 The mean ACh sensitivity of surface fibers of rat extensor muscle before (○), 1 day (●) and 2 days (□) after sectioning of the motor nerve. The abscissa shows the distance from the endplate region (from Albuquerque and McIsaac [243]).

extrajunctional membrane. Both muscles become highly sensitive to ACh, but not as much as at the endplate region. The denervation-induced ACh sensitivity thus occurs in patches of high and low receptor density [243]. Between 14 and 45 days after denervation, ACh sensitivities of both muscles decrease.

Increase in ACh receptors upon denervation also occurs in neurons, as in those from the frog parasympathetic ganglion, with ACh receptors expanding to cover the entire neuron after the incoming axons are cut [255]. In the rabbit superior cervical ganglion cells, the increase in ACh sensitivity, resulting from denervation, was attributed to an increase in muscarinic receptors, not nicotinic ones [256].

Effects similar to denervation can be obtained by administration of receptor inhibitors such as α-BGT or curare, which eliminate transmission [257]. An increase in ACh receptors also occurs when eserine is used to inhibit cholinesterases in innervated muscles [189]. The spread of receptors is detectable by electrophysiological means, as an increase in ACh sensitivity, and in binding of $[^{125}I]$α-BGT [257, 258] and these receptors are not related to the tetrodotoxin-resistant sodium channels [259]. However, the increase in α-BGT binding is much less than the increase in ACh sensitivity [225] which represents a point still unresolved.

Junctional (at the endplate) and extrajunctional (outside the endplate) ACh receptors bind ACh and respond by depolarization of the muscle membrane. Furthermore, ACh receptors purified from rat diaphragm muscle and those purified from denervated diaphragm have similar molecular weight and both bind concanavalin A (indicative of presence of sugars). However, the affinity for binding of *d*-tubocurarine is about 10-fold lower for the extrajunctional receptors, and their isoelectric point is $\simeq 0.15$ units higher [260]. Electrophysiological studies on the denervated extrajunctional and junctional ACh receptors of skeletal muscles from mice and rats demonstrated that α-BGT bound more reversibly to the extrajunctional receptors and hexamethonium was ineffective in blocking this binding [166]. *d*-Tubocurarine was 20-fold less potent as an inhibitor of ACh response in denervated than in innervated muscle [189]. Extrajunctional receptors had a faster turnover rate (see Section 5.6.4) and the open time for extrajunctional channels was 3–5-fold longer than junctional channels [261]. It is not known whether the change in open time for channels is due to a difference in receptor or in the channel molecule itself or both. Additional differences between junctional and extrajunctional ACh receptors were noted from immunologic studies. Immunoglobulins from patients with myasthenia gravis blocked binding of α-BGT to rat extrajunctional solubilized ACh receptors, but not to junctional ones [153]. Conversely, immunoglobulins from rats injected with pure *Torpedo* ACh receptors bound selectively to rat and human junctional receptors and not to extrajunctional ones [262]. It is interesting to know whether the same or separate genes code for extrajunctional and junctional receptors.

5.6.3 Trophic factors and synapse formation

Not only do neurotransmitters move from neurons to neurons or other cells across synapses; other molecules also do. In fact this transfer of molecules goes in both directions and is important for cell viability. Neurotrophic effects have been defined by Guth [263] as the 'interactions between nerve and other cells which initiate or control molecular modification in the other cell', and this occurs by means of diffusible substances called neurotrophic factors. An example of established neurotrophic effects is the degeneration of taste buds when their nerve supply is interrupted and their reappearance after nerve regeneration [264].

Albuquerque *et al.* [265] found that, upon denervation, muscle contraction ceased first, then resting membrane potential decreased and finally reductions in MEPPs occurred. The closer to the endplate the nerve was cut, the sooner was the change in resting membrane potential and the increase in ACh sensitivity. Also membrane potential increased first during re-innervation of skeletal muscle following nerve crush, then the ACh sensitivity returned to normal [266]. Thus, a neurotrophic mechanism has to be postulated because, when the nerve is severed, there is no longer neuromuscular transmission or muscle activity [266].

ACh receptors are synthesized in muscle cells away from any neuronal input, but they are also influenced neurotrophically. Extrajunctional ACh receptors exhibit differences in their pharmacology and biochemistry (see Section 5.6.2). It is difficult to determine what causes the increase in ACh receptors that occurs upon denervation since several changes occur as a result, such as cessation of transmission and absence of ACh and other chemicals that may be released by the neuron. It has been shown that blockade of either axoplasmic flow in the neuron (e.g. with colchicine or vinblastine [267]), or ACh release (e.g. indirectly by blocking impulse conduction with tetrodotoxin [268]), causes increased extrajunctional ACh sensitivity, and if both occur, as when a nerve is severed, then the increase may be even greater. Also, when the ACh receptors of skeletal muscle are blocked with α-BGT, or botulinum toxin blocks ACh release, effects similar to those of denervation are obtained [257, 269]. On the other hand, the reverse events (i.e. localization of ACh receptors to the site of contact between a neuron and a muscle cell) were found to be independent of release of ACh from the neuron and of the presence of ACh receptors [241], but the effect of axoplasmic flow was not determined.

Another factor that is important in regulating the distribution of ACh receptors is muscle activity. In fact, electrical stimulation of denervated muscles has been shown to prevent or partially reverse the extrajunctional hypersensitivity to ACh [270–272]. In addition, inhibition of the spontaneous contraction of chick muscle fibers in culture increased the concentration of ACh receptors by 4–8-fold, and electrical stimulation decreased it, possibly by partial suppression of receptor synthesis and not inactivation of existing ACh receptors [270]. The increase in extrajunctional ACh receptors in rat skeletal muscles resulting from muscle disuse was not as great as that resulting from denervation, which again suggests that, in addition to muscle activity, neurotrophic factors play a role [273].

Not only is the ACh receptor regulated neurotrophically, but it appears to regulate pre-synaptic components. Chick embryos injected with cobrotoxin had marked reduction in growth, number of endplates, differentiation of spinal roots, sciatic nerve and ganglia [274]. The molecular mechanism involved in such trophic effects is still unknown.

It is important to know if there are cues that direct an axon to a certain location in the muscle cell to form a synapse. Neurons and skeletal muscle fragments [275, 276] or myotubes [277–280] have been shown to form synapses in cultures. It appears that most muscle cells have the same specificity for synapse formation, and that formation of a synapse between neuron and muscle cell is not dependent on interactions between complementary molecules on both cells that code for different synaptic connections [281].

5.6.4 Synthesis and degradation of ACh receptors

The appearance of new ACh receptors in developing chick myotubes or denervated mammalian muscles appears to be due to *de novo* synthesis and not unmasking of pre-existing receptors, because of its dependence on RNA and protein synthesis [282–284] and the incorporation of [^{35}S] methionine into new receptors of denervated rat hemidiaphragm [285] and myoblasts [286, 287]. This is also confirmed by the finding that incubation of chick myotubes in [^{2}H] or [^{13}C] substituted amino acids produces new ACh receptors that have increased buoyancy due to incorporation of these amino acids [288]. There is also some evidence from kinetic analysis and subcellular fractionation, that ACh receptors are incorporated into some internal membrane system immediately after synthesis, and that this pool can supply the cell surface with receptors for 2–3 h in the absence of protein synthesis [289].

In developing muscles, ACh receptors must be synthesized at a rate that is faster than their degradation so that the final outcome is accumulation of receptors. [^{125}I] α-BGT was utilized as a specific marker of the ACh receptor in studies of its degradation, whether in rat or chick myotubes in cell culture by Devreotes and Fambrough [289] or in rat *in vivo* by Berg and Hall [290]. In the latter study degradation of ACh receptors was blocked by protein synthesis inhibitors. From these studies and others [287, 291], we may conclude that junctional ACh receptors are metabolically very stable (half-life of toxin loss is 6 days) while extrajunctional ACh receptors and those of myotubes in culture are much more labile (half-life of one day or less).

It is quite apparent that although a long road is still ahead, utilization of various cell lines in studies of ACh receptors is yielding most interesting and exciting data. These systems provide opportunities to examine the effect of one neuron on the ACh receptor of another, mediation of nerve–muscle trophic interactions and to study the regulation, distribution and turnover of ACh receptors.

5.7 THE ACh RECEPTOR AND DISEASED STATES

5.7.1 Myasthenia gravis

Myasthenia gravis is a human disease which is characterized by abnormality in neuromuscular transmission. There is reduced amplitude of MEPP, progressive reduction of muscle action potential response to repetitive nerve stimulation and improvement after administration of anticholinesterases. Until recently, it was not certain whether the disease was due to a decrease in the amount of ACh released by the motor nerve terminals or in the action of ACh at the postsynaptic membrane. In 1960, Simpson [292] postulated an autoimmune etiology for the disease and proposed that the defect was postsynaptic. This was confirmed by elaborate studies of Albuquerque *et al.* [293] on muscle biopsies from myasthenic patients where they showed significant destruction of postsynaptic membrane with a widening of the synaptic gap and reduction in ACh sensitivity, suggesting an abnormality in the ACh receptor and/or its ICM. It was also found that the number of α-BGT binding molecules (i.e. ACh receptors) in myasthenic endplates was reduced. Antibodies were detected in most myasthenics, which blocked α-BGT binding to normal human biopsies or solubilized denervated rat ACh receptor [294–297] and precipitated [^{125}I]α-BGT bound to human muscle ACh receptors [149], thus confirming that myasthenia gravis is an autoimmune disease. However, since there was no close correlation between antibody titer and disease intensity, it was suggested that the disease process involved more than the binding of antibody to ACh receptor [149]. Cellular immunity also appears to be involved [298].

The pioneering work of Patrick and Lindstrom [299], followed by many others [300–303], demonstrated that animals immunized with ACh receptors purified from electric organs developed muscular weakness that could reach flaccid paralysis and rapid death as in rabbits. Muscle weakness was characterized by reduced MEPP amplitudes, temporary improvement with anticholinesterases, and increased sensitivity to *d*-tubocurarine [304, 305]. Antisera from the immunized animals blocked carbamylcholine-induced depolarization in isolated electroplax and cultured rat myoblasts [299, 301], and reduced ACh receptor numbers in normal rats partly by targeting postsynaptic membrane for destruction by phagocytic cells. The antisera also blocked [^{3}H] ACh binding to the *Torpedo* membrane-bound, solubilized and purified ACh receptors [302]. Sanders *et al.* [306] found that removal of the thymus (a treatment that helps many myasthenics) prior to inoculation with receptor slowed the development of alterations in MEPP. In addition, myasthenia-like symptoms appeared in rats which received daily injections of immunoglobulins from myasthenic patients [307]. Cellular immunity also appears to be involved in the animal model [150].

The ACh sensitivity of denervated muscle in rats immunized with purified *Electrophorus* ACh receptors was reduced, as was that of the unimmunized rats after exposure to antisera from rats immunized with the receptor [308]. This suggests that the defect produced is in the ACh receptor of denervated muscle. It is in harmony with the assumption that the preparation used to immunize these animals is indeed a purified ACh receptor.

The accepted hypothesis is that immunization with purified nicotinic ACh receptor leads to a breakdown in the immunological tolerance of the animal for its own nicotinic neuromuscular receptor, with subsequent humoral and cellular reactions against the animal's own skeletal muscle ACh receptors. This animal model for myasthenia gravis has been studied extensively since its discovery in 1972 and, in every aspect, has proved to be similar to myasthenia gravis. It is producing a great deal of excitement among neurologists because it represents the first promising model for this neurologic disease.

5.7.2 Huntington's chorea

Huntington's chorea is a hereditary disorder that is characterized by involuntary choreic movements and psychotic behavior [309]. Two models have been produced for this disease by administration of *l*-dopa [310] and recently by 3-methoxytyramine [311]. There is implication of some disturbance of the dopaminergic system since dopamine antagonists also alleviate some of the symptoms [312]. Post-mortem brain samples (corpus striatum) had reduced activity of the γ-aminobutyric acid synthesizing enzyme and choline acetyltransferase. However, γ-aminobutyric acid receptors and β-adrenergic receptors were not reduced in caudate nucleus, although serotonin and muscarinic receptors were greatly reduced [312]. The caudate nucleus has the highest concentration of only the muscarinic type ACh receptors [313–315]. Since serotonin levels in patients were unchanged and tryptophan (the precursor of serotonin) had no influence on symptoms, it was suggested that reduced muscarinic receptors was possibly a cause for the disease. Hiley and Bird [316] reported a greatly reduced concentration of muscarinic ACh receptors in the caudate nucleus of choreic brain relative to control samples, as well as in several other areas of the brain, ranging from 35% reduction in the hippocampus to 100% in the globus pallidus [313], but no change in cortex receptors.

If the cause of this disease is indeed a reduction in muscarinic receptors, we might ask what causes the loss in receptors. Is it possible that it is due to an autoimmune response analogous to myasthenia gravis? If it is, a search for anti-muscarinic ACh receptor antibodies in the blood of choreic patients might be rewarding.

5.7.3 Leukemia

Although there is no direct correlation between ACh receptors and leukemia, circumstantial evidence points to a small probability that ACh receptors may be involved. Bone marrow stem cells possess the property of self-renewal and also give rise to cells that are committed to differentiate into a particular cell, e.g. erythrocytes, macrophages, granulocytes. If bone marrow cells are transplanted into a heavily irradiated mouse they form colonies in the spleen [317]. Under conditions that require increased numbers of differentiated bone marrow cells, the stem cells enter into a cell cycle, thus becoming more sensitive to [^{3}H] thymidine and to inhibitors of DNA synthesis. Therefore, cells that entered into cell cycle and multiplication and were exposed to [^{3}H] thymidine formed fewer colonies in the spleen of a heavily irradiated mouse [318]. Byron [319] discovered that several chemicals induced the bone marrow stem cells into cycle, examples being isoproterenol, ACh (with neostigmine) and carbamylcholine. The cholinergic agonists were effective at the extremely low concentrations of 10^{-14} – 10^{-12} M and their action was blocked with 10^{-6} M *d*-tubocurarine (Fig. 5.14). It was suggested that cGMP might mediate this action, since dibutyl cGMP had the same effect as ACh. β-adrenergic blockade did not antagonize cholinergic agonists. These data led Byron to many important questions concerning

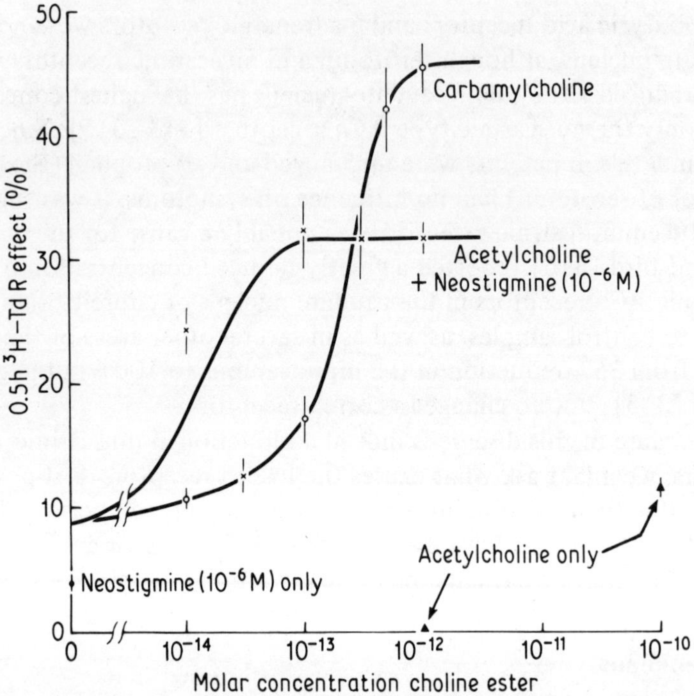

Fig. 5.14 Percentage loss of spleen colony-forming units due to [^{3}H] thymidine in bone marrow cultures treated with carbamylcholine, ACh, neostigmine or with both neostigmine and ACh (from Byron [319]).

the possibility of these mechanisms acting at the physiologic level to regulate the proliferation of the hemopoietic stem cell and whether the nerve endings that terminate close to bone marrow cells or other cells produce the triggering chemicals.

It is very interesting to know why an ACh receptor would have such a high affinity for agonists, what effect therapeutic cholinergic drugs might have on the proliferation of these stem cells, what role, if any, would the ACh receptor play in the leukemogenic process and whether it is affected by carcinogens.

5.8 CONCLUSION

In the last eigh years, the ACh receptor has developed from a hypothetical entity into a membrane protein that can be visualized, whose molecular weight, subunit structure, composition and interaction with drugs are better known. It has become involved in a human neurologic disorder, and we may speculate as to other involvements. Yet we are still far away from total understanding of this molecule and the varieties of shapes it exists in or functions it has. There are tissues where ACh is present, but in which ACh receptors have not been discovered, such as the placenta. On the other hand, the ACh receptor is detected in tissues, such as bone marrow stem cells, where its function is still unknown.

REFERENCES

1. Bernard, C. (1857), *Leçons sur les Effets des Substances Toxiques et Médicamenteuses,* Baillière, Paris,
2. Langley, J.N. and Dickinson, W.L. (1889), *Proc. Roy. Soc. (Lond.), B.* **46**, 423–431.
3. Langley, J.N. (1906), *J. Physiol.,* **20**, 223–246.
4. Dale, H.H. (1914), *J. Pharmac. exp. Ther.,* **6**, 147–190.
5. Loewi, O. (1921), *Arch. Ges. Physiol. Pflügers,* **189**, 239–242.
6. Loewi, O. and Navratil, E. (1926), *Arch. Ges. Physiol. Pflügers,* **214**, 678–688.
7. Rall, T.W., Sutherland, W.W. and Berthet, J. (1957), *J. biol. Chem.,* **224**, 463–475.
8. Rall, T.W. and Sutherland, E.W. (1958), *J. biol. Chem.,* **232**, 1065–1076.
9. Dubois-Raymond, E. (1887), In: *Translations of Foreign Biological Memoirs. I. On the Physiology of Nerve, of Muscle and of the Electric Organ.* (J. Burdon-Sanderson, ed.), Oxford University Press, London, pp. 367–413.
10. Nachmansohn, D. (1955), *Harvey Lect.,* **49**, 57–99.
11. Feldberg, W. and Fressard, A. (1942), *J. Physiol.,* **101**, 200–216.
12. Keynes, R.D. and Martins-Ferreira, H. (1953), *J. Physiol.,* **119**, 315–351.
13. Bennett, M.V.L. (1970), *Ann. Rev. Physiol.,* **32**, 471–524.
14. Schoffeniels, E. and Nachmansohn, D. (1957), *Biochim. biophys. Acta,* **26**, 1–15.
15. Gerschenfeld, H.M. (1973), *Physiol. Rev.,* **53**, 1–119.
16. Purves, R.D. (1976), *Nature,* **61**, 149–151.
17. Albuquerque, E.X., Barnard, E.A., Porter, C.W. and Warnick, J.E. (1974), *Proc. natn. Acad. Sci. U.S.A.,* **71**, 2818–2822.
18. Fertuck, H.C. and Salpeter, M.M. (1974), *Proc. natn. Acad. Sci. U.S.A.,* **71**, 1376–1378.
19. Heuser, J.E., Reese, T.S. and Landis, D.M.D. (1974), *J. Neurocyt.,* **3**, 109–131.
20. Blankenship, J.E., Wachtel, H. and Kandel, E.R. (1971), *J. Neurophys.,* **34**, 76–92.

21. Fatt, P. and Katz, B. (1951), *J. Physiol.*, **115**, 320–370.
22. Katz, B. and Miledi, R. (1970), *Nature*, **226**, 962–963.
23. Katz, B. and Miledi, R. (1971), *Nature New Biol.*, **232**, 124–126.
24. Katz, B. and Miledi, R. (1972), *J. Physiol.*, **224**, 665–699.
25, Katz, B. and Miledi, R. (1975), *Proc. Roy. Soc., (Lond.) B.*, **192**, 27–38.
26. Katz, B. and Miledi, R. (1973), *J. Physiol.*, **230**, 707–717.
27. Salpeter, M.M. and Eldefrawi, M.E. (1973), *J. Histochem. Cytochem.*, **21**, 769–778.
28. Magleby, K.L. and Stevens, C.F. (1972), *J. Physiol.*, **223**, 151–171.
29. Magleby, K.L. and Stevens, C.F. (1972), *J. Physiol.*, **223**, 173–197.
30. Anderson, C.R. and Stevens, C.F. (1973), *J. Physiol.*, **235**, 655–691.
31. Stevens, C.F. (1974), In: *Synaptic Transmission and Neuronal Interactions* (M.V.L. Bennett, ed.) Raven Press, New York, pp. 45–58.
32. Hubbard, J.I. and Schmidt, R.F. (1963), *J. Physiol.*, **166**, 145–167.
33. Tosaka, T., Chichibu, S. and Libet, B. (1968), *J. Neurophysiol.*, **31**, 396–409.
34. Bradley, P.B. and Wolstencroft, J.H. (1965), *Br. Med. Bull.*, **12**, 15–18.
35. McCance, I., Phillis, J.W., Tebecis, A.K. and Westerman, R.A. (1968), *Br. J. Pharmac. Chemother.*, **32**, 652–662.
36. Chothia, C. (1970), *Nature*, **225**, 36–38.
37. Chothia, C. and Pauling, P. (1970), *Proc. natn. Acad. Sci. U.S.A.*, **65**, 477–482.
38. Pauling, P. and Petcher, T.J. (1972), *Nature New Biol.*, **236**, 112–113.
39. Waser, P.G. (1962), *Pharmacol. Rev.*, **13**, 465–515.
40. Green, L. (1976), *Brain Res.*, **111**, 135–145.
41. Moore, W. Mc. and Brady, R.N. (1976), *Biochim. biophys. Acta*, **444**, 252–260.
42. Wang, G.-K. and Schmidt, J. (1976), *Brain. Res.*, **114**, 524–529.
43. Brown, D.A., Garthwaite, J., Hyashi, E. and Yamada, S. (1976), *Br. J. Pharmac.*, **58**, 157–159.
44. Faeder, I.R., O'Brien, R.D. and Salpeter, M.M. (1970), *J. exp. Zool.*, **173**, 203–214.
45. Lunt, G.G. (1975), In: *Insect Biochemistry and Function* (D. J. Candy and B.A. Kilby, eds.) Chapman and Hall, London, pp. 283–306.
46. Colhoun, E.H. (1963), *Adv. Insect Physiol.*, **1**, 1–46.
47. Smallman, B.N. and Fischer, R.W. (1961), *Can. J. Biochem. Physiol.*, **36**, 575–586.
48. Roeder, K.D. and Roeder, S. (1939), *J. Cell. Comp. Physiol.*, **14**, 1–12.
49. Roeder, K.D. (1948), *Bull. Johns Hopkins Hosp.*, **83**, 587–600.
50. Shankland, D.J., Rose, J.A. and Donniger, C. (1971), *J. Neurobiol.*, **2**, 247–262.
51. Flattum, R.F. and Shankland, D.L. (1971), *Comp. Gen. Pharm.*, **2**, 159–167.
52. Eldefrawi, A.T. and O'Brien, R.D. (1970), *J. Neurochem.*, **17**, 1287–1293.
53. Eldefrawi, M.E., Eldefrawi, A.T. and O'Brien, R.D. (1970), *Mol. Pharmacol.*, **7**, 104–110.

54. Aziz, S.A. and Eldefrawi, M.E. (1973), *Pestic. Biochem. Physiol.*, **3**, 168–174.
55. Barnard, E.A., Dolly, J.O., Porter, C.W. and Albuquerque, E.X. (1975), *Exp. Neurol.*, **48**, 1–28.
56. Negrette, J., Del Castillo, J., Escobar, I. and Yankelovich, G. (1972), *Nature New Biol.*, **235**, 158–159.
57. Hubbard, J.I. (1974), In: *The Peripheral Nervous System* (J. I. Hubbard, ed.) Plenum Press, New York, pp. 151–180.
58. Potter, L.T. (1970), *J. Physiol.*, **206**, 145–166.
59. Katz, B. and Miledi, R. (1973), *J. Physiol.*, **231**, 549–574.
60. Albuquerque, E.X., Barnard, E.A., Jansson, S.-E. and Wieckowski, J. (1973), *Life Sci.*, **12**, 545–552.
61. Triggle, C.R. and Triggle, D.J. (1976), *J. Physiol.*, **254**, 39–54.
62. Nastuk, W.L. and Hodgkin, A.L. (1950), *J. Cell Comp. Physiol.*, **35**, 39–73.
63. Schoffeniels, E. (1957), *Biochim. biophys. Acta*, **26**, 585–596.
64. Moreau, M. and Changeux, J.-P. (1976), *J. mol. Biol.*, **106**, 457–467.
65. Ferrendelli, J.A., Kinscherf, D.A. and Chang, M.M. (1973), *Mol. Pharmacol.*, **9**, 445–454.
66. Steiner, A.L., Parker, C.W. and Kipnis, D.M. (1972), *J. biol. Chem.*, **247**, 1106–1113.
67. George, W.J., Polson, J.B., O'Toole, A.G. and Goldberg, N.D. (1970), *Proc. natn. Acad. Sci. U.S.A.*, **66**, 398–403.
68. Eldefrawi, M.E., Britten, A.G. and Eldefrawi, A.T. (1971), *Science*, **173**, 338–340.
69. O'Brien, R.D., Gilmour, L.P. and Eldefrawi, M.E. (1970), *Proc. natn. Acad. Sci. U.S.A.*, **65**, 438–445.
70. Miledi, R., Molinoff, P. and Potter, L.T. (1971), *Nature*, **229**, 554–557.
71. Karlin, A., Prives, J., Deal, W. and Winnik, M. (1970), In: *Molecular Properties of Drug Receptors.* (R. Porter and M. O'Connor, eds.) Ciba Foundation Symposium, J. and A. Churchill, London, pp. 247–261.
72. Eldefrawi, M.E., Eldefrawi, A.T. and O'Brien, R.D. (1971), *Proc. natn. Acad. Sci. U.S.A.*, **68**, 1047–1050.
73. Suszkiw, J.B. (1973), *Biochim. biophys. Acta*, **318**, 69–77.
74. Raftery, M.A., Schmidt, J., Clark, D.G. and Wolcott, R.G. (1971), *Biochem. biophys. Res. Comm.*, **45**, 1622–1629.
75. Schleifer, L.S. and Eldefrawi, M.E. (1974), *Neuropharmacol.*, **13**, 53–63.
76. Hiley, C.R. and Burgen, A.S.V. (1974), *J. Neurochem.*, **22**, 159.
77. Salvaterra, P.M. and Moore, W.J. (1973), *Biochem. biophys. Res. Comm.*, **55**, 1311–1318.
78. O'Brien, R.D. and Gilmour, L.P. (1969), *Proc. natn. Acad. Sci. U.S.A.*, **63**, 496–503.
79. Lindstrom, J. and Patrick, J. (1974), In: *Synaptic Transmission and Neuronal Interactions.* (M. V. L. Bennett, ed.) Raven Press, New York, pp. 191–216.
80. Eldefrawi, A.T. (1976), In: *Insecticide Biochemistry and Physiology.* (C.F. Wilkinson, ed.) Plenum Publishing Corp., New York. pp. 297–326.

81. Donner, D.B., Fu, J.-J. and Hess, G.P. (1976), *Analyt. Biochem.*, **75**, 454–463.
82. Franklin, G.I. and Potter, L.T. (1972), *FEBS Letters*, **28**, 101–106.
83. Schmidt, J. and Raftery, M.A. (1973), *Analyt. Biochem.*, **52**, 349–354.
84. Almon, R.R., Andrew, C.G. and Appel, S.H. (1974), *Biochemistry*, **13**, 5522–5528.
85. Kohanski, R.A., Andrews, J.P., Wins, P., Eldefrawi, M.E. and Hess, G.P. *Analyt. Biochem.*, (in press).
86. Kasai, M. and Changeux, J.-P. (1971), *J. memb. Biol.*, **6**, 1–23.
87. Kasai, M. and Changeux, J.-P. (1971), *J. memb. Biol.*, **6**, 24–57.
88. Popot, J.-L., Sugiyama, H. and Changeux, J.-P. (1976), *J. mol. Biol.*, **106**, 469–483.
89. Hess, G.P., Andrews, J.P., Struve, G.E. and Coombs, S.E. (1975), *Proc. natn. Acad. Sci. U.S.A.*, **72**, 4371–4375.
90. Flanagan, S.D., Taylor, P. and Barondes, S.H. (1975), *Nature*, **254**, 441–443.
91. Changeux, J.-P., Kasai, M., Huchet, M. and Meunier, J.-C. (1970), *C.R. Acad. Sci. Paris Ser. D*, **270**, 2864–2867.
92. Raftery, M.A., Schmidt, J. and Clark, D.G. (1972), *Arch. Biochem. Biophys.*, **152**, 882–886.
93. Eldefrawi, M.E., Eldefrawi, A.T., Seifert, S. and O'Brien, R.D. (1972), *Arch. Biochem. Biophys.*, **150**, 210–218.
94. Eldefrawi, M.E. and Eldefrawi, A.T. (1972), *Proc. natn. Acad. Sci. U.S.A.*, **69**, 1176–1780.
95. Cuatrecasas, P. (1970), *J. biol. Chem.*, **245**, 3059–3065.
96. Karlsson, E., Heilbronn, E. and Widlund, L. (1972), *FEBS Letters*, **28**, 107–111.
97. Klett, R.P., Fulpius, B.W., Cooper, D., Smith, M., Reich, E. and Possani, L. (1973), *J. biol. Chem.*, **248**, 6841–6853.
98. Eldefrawi, M.E. and Eldefrawi, A.T. (1973), *Arch. Biochem. Biophys.*, **159**, 362–373.
99. Biesecker, G. (1973), *Biochemistry*, **12**, 4403–4409.
100. Ong, D.E. and Brady, R.N. (1974), *Biochemistry*, **13**, 2822–2827.
101. Olsen, R.W., Meunier, J.-C. and Changeux, J.-P. (1972), *FEBS Letters*, **28**, 96–100.
102. Schmidt, J. and Raftery, M.A. (1973), *Biochemistry*, **12**, 852–856.
103. Chang, H.W. (1974), *Proc. natn. Acad. Sci. U.S.A.*, **71**, 2113–2117.
104. Michaelson, D., Vandlen, R., Bode, J., Moody, T., Schmidt, J. and Raftery, M.A. (1974), *Arch. Biochem. Biophys.*, **165**, 796–804.
105. Eldefrawi, M.E., Eldefrawi, A.T. and Wilson, D.B. (1975), *Biochemistry*, **14**, 4304–4309.
106. Rübsamen, H., Eldefrawi, A.T., Eldefrawi, M.E. and Hess, G.P., *Biochemistry*, (in press).
107. de Robertis, E. (1974), In: *Neurochemistry of Cholinergic Receptors.* (E. de Robertis and J. Schacht, eds.) Raven Press, New York, pp. 63–84.

108. Barrantes, F.J., Changeux, J.-P., Lunt, G.C. and Sobel, A. (1975), *Nature,* **256**, 325–327.
109. Bradley, R.J., Howell, J.G., Romine, W.O., Carl, G.F. and Kemp, G.E. (1976), *Biochem. biophys. Res. Comm.,* **68**, 577–584.
110. Slavaterra, P.M. and Mahler, R. (1976), *J. biol. Chem.,* **251**, 6327–6334.
111. Beld, A.J. and Ariens, E.J. (1974), *Eur. J. Pharm.,* **25**, 203–209.
112. Eldefrawi, A.T., Eldefrawi, M.E. and Mansour, N.A. In: *Pesticide and Venom Neurotoxicity* (D.L. Shankland, R.F. Flattum, R.M. Hollingworth and T. Smyth, Jr., eds.) Plenum Publishing Corp., New York (in press).
113. Kato, G. and Tattrie, B. (1974), *FEBS Letters,* **48**, 26–31.
114. Moore, W.M., Holladay, L.A., Puett, D. and Brady, R.N. (1974), *FEBS Letters,* **45**, 145–149.
115. Meunier, J.-C., Sealock, R., Olsen, R. and Changeux, J.-P. (1974), *Eur. J. Biochem.,* **45**, 371–394.
116. Heilbronn, E. and Mattson, C. (1974), *J. Neurochem.,* **22**, 315–317.
117. Eldefrawi, M.E. and Eldefrawi, A.T. (1975), *Croatica Chem. Acta,* **47**, 425–438.
118. Barrantes, F.J. (1975), *Biochim. biophys. Acta,* **62**, 407–414.
119. Brisson, A.D., Scandella, C.J., Bienvenue, A., Devaux, P.F., Cohen, J.B. and Changeux, J.-P. (1975), *Proc. natn. Acad. Sci. U.S.A.,* **72**, 1087–1091.
120. Brand, L. and Withholt, B. (1967), *Meth. Enzymol.,* **11**, 776–856.
121. Eldefrawi, M.E., Eldefrawi, A.T., Penfield, L.A., O'Brien, R.D. and Van Campen, D. (1974), *Life Sci.,* **16**, 925–936.
122. Chang, H.W. and Neumann, E. (1976), *Proc. natn. Acad. Sci. U.S.A.,* **73**, 3364–3368.
123. Edelstein, S.E., Beyer, W.B., Eldefrawi, A.T. and Eldefrawi, M.E. (1975), *J. biol. Chem.,* **250**, 6101–6106.
124. Meunier, J.-C., Olsen, R.W., Menez, A., Fromageot, P., Boquet, P. and Changeux, J.-P. (1972), *Biochem.,* **11**, 1200–1210.
125. Hucho, F. and Changeux, J.-P. (1973), *FEBS Letters,* **38**, 11–15.
126. Martinez-Carrion, M., Sator, V. and Raftery, M.A. (1975), *Biochem. biophys. Res. Comm.,* **65**, 129–137.
127. Gibson, R.E., O'Brien, R.D., Edelstein, S.J. and Thompson, W.R. (1976), *Biochemistry,* **15**, 2377–2383.
128. Carroll, R.C., Eldefrawi, M.E. and Edelstein, S.J. (1973), *Biochem. biophys. Res. Comm.,* **58**, 864–872.
129. Patrick, J., Boulter, J. and O'Brien, J.C. (1975), *Biochem. biophys. Res. Comm.,* **64**, 219–225.
130. Hucho, F., Layer, P., Kiefer, H.R. and Bandini, G. (1976), *Proc. natn. Acad. Sci. U.S.A.,* **73**, 2624–2628.
131. Eldefrawi, M.E., Eldefrawi, A.T. and Shamoo, A.E. (1975), *Ann. N.Y. Acad. Sci.,* **264**, 183–202.
132. Rübsamen, H., Montgomery, M., Hess, G.P., Eldefrawi, A.T. and Eldefrawi, M.E. (1976), *Biochem. biophys. Res. Comm.,* **70**, 1020–1027.

133. Cartaud, J., Benedetti, E.L., Cohen, J.B., Meunier, J.-C. and Changeux, J.-P. (1973), *FEBS Letters,* **33**, 109–113.
134. Nickel, E. and Potter, L.T. (1973), *Brain Res.,* **57**, 508–517.
135. Benedetti, E.L. and Emmelot, P. (1968), *J. Cell Biol.,* **38**, 15–24.
136. Zampighi, G. and Robertson, J.D. (1973), *J. Cell Biol.,* **56**, 92–105.
137. Dupont, Y., Cohen, J.B. and Changeux, J.-P. (1974), *FEBS Letters,* **40**, 130–133.
138. Paton, W.D.M. (1961), *Proc. Roy. Soc. (Lond.) B.,* **154**, 21–69.
139. Stephenson, R.P. (1956), *Br. J. Pharmac. Chemother.*, **11**, 379–393.
140. Ariens, E.J., Simonis, A.M. and van Rossum, J.M. (1964), In: *Molecular Pharmacology* (E.J. Ariens, ed.) Academic Press, New York., pp. 287–393.
141. Furchgott, R.F. (1966), *Adv. Drug Res.,* **3**, 21.
142. Eldefrawi, M.E. (1974), In: *The Peripheral Nervous System.* (J. I. Hubbard, ed.) Plenum Press, New York, pp. 181–200.
143. Bulger, J.E. and Hess, G.P. (1973), *Biochem. biophys. Res. Comm.,* **54**, 677–684.
144. Hess, G.P., Bulger, J.E., Fu, J.-J. L., Hindy, E.F. and Silberstein, R.J. (1975), *Biochem. biophys. Res. Comm.,* **64**, 1018–1027.
145. Shamoo, A.E., MacLennan, D.H. and Eldefrawi, M.E. (1976), *Chem. Biol. Interactions,* **12**, 41–52.
146. Eldefrawi, M.E., Mansour, N.A. and Eldefrawi, A.T., In: *Membrane Toxicity* (A.E. Shamoo, ed.) Plenum Press, New York, (in press).
147. Chao, Y., Vandlen, R.L. and Raftery, M.A. (1975), *Biochem. biophys. Res. Comm.,* **63**, 300–307.
148. Gibson, R.E. (1976), *Biochemistry,* **15**, 3890–3901.
149. Lindstrom, J.M., Seybold, M.E., Lennon, V.A., Whittingham, S. and Duane, D.D. (1976), *Neurology,* **26**, 1054–1059.
150. Lennon, V.A., Lindstrom, J.M. and Seybold, M.E. (1976), *Ann. N.Y. Acad. Sci.,* **274**, 283–299.
151. Fulpius, B.W., Zurn, A.D., Granato, D.A. and Leder, R.M. (1976), *Ann. N.Y. Acad. Sci.,* **274**, 116–129.
152. Lindstrom, J. (1976), *J. Supramol. Struct.,* **4**, 389–403.
153. Almon, R.R. and Appel, S.H. (1976), *J. Neurochem.,* **27**, 1513–1519.
154. Ross, D.H. and Triggle, D.J. (1974), *Can. J. Physiol. Pharmacol.,* **52**, 323–333.
155. Gupta, S.K., Moran, J.F. and Triggle, D.J. (1976), *Mol. Pharmacol.,* **12**, 1019–1026.
156. Rang, H.P. (1973), *Nature,* **231**, 91–96.
157. Changeux, J.-P., Podleski, T. and Meunier, J.-C. (1969), *J. gen. Physiol.,* **54**, 225–244.
158. Eldefrawi, M.E. and Eldefrawi, A.T. (1973), *Biochem. Pharmacol.,* **22**, 3145–3150.
159. Meunier, J.-C., Sugiyama, H., Cartaud, J., Sealock, R. and Changeux, J.-P. (1973), *Brain Res.,* **62**, 307–315.

160. O'Brien, R.D. and Gibson, R.E. (1975), *Arch. Biochem. Biophys.*, **169**, 458–463.
161. Monod, J., Wyman, J. and Changeux, J.-P. (1965), *J. mol. Biol.*, **12**, 88–118.
162. Edelstein, S.J. (1972), *Biochem. biophys. Res. Comm.*, **48**, 1160–1165.
163. Koshland, D.E., Jr., Némethy, G. and Filmer, D. (1966), *Biochemistry*, **5**, 365–385.
164. Weber, M. and Changeux, J.-P. (1974), *Mol. Pharmacol.*, **10**, 1–14.
165. Weber, M. and Changeux, J.-P. (1974), *Mol. Pharmacol.*, **10**, 15–34.
166. Sarvey, J.M., Albuquerque, E.X., Eldefrawi, A.T. and Eldefrawi, M.E., *Memb. Biochem.*, (in press).
167. Barrantes, F.J. (1976), *Biochem. biophys. Res. Comm.*, **72**, 429–488.
168. Bonner, R., Barrantes, F.J. and Jovin, T.M. (1976), *Nature*, **263**, 429–431.
169. Grünhagen, H.-H. and Changeux, J.-P. (1976), *J. mol. Biol.*, **106**, 517–535.
170. Cohen, J.B. and Changeux, J.-P. (1973), *Biochemistry*, **12**, 4855–4864.
171. Rübsamen, H., Hess, H.P., Eldefrawi, A.T. and Eldefrawi, M.E. (1976), *Biochem. biophys. Res. Comm.*, **68**, 56–63.
172. Frankenhaeuser, B. and Hodgkin, A.L. (1957), *J. Physiol.*, **137**, 218–244.
173. Hurwitz, L. and Suria, A. (1971), *Ann. Rev. Pharmacol.*, **11**, 303–326.
174. Chang, K.-J. and Triggle, D.J. (1973), *J. Theor. Biol.*, **40**, 125–154.
175. Goodman, F.R. and Weiss, G.B. (1971), *Am. J. Physiol.*, **220**, 759–766.
176. Martinez-Carrion, M. and Raftery, M.A. (1973), *Biochem. biophys. Res. Comm.*, **55**, 1156–1164.
177. Magnusson, S., Sottrup-Jensen, L., Petersen, T.E., Morris, H. R. and Dell, A. (1974), *FEBS Letters*, **44**, 189 193.
178. Nelsestuen, G.L., Broderius, M., Zytkovicz, T.H. and Howard, J. (1975), *Biochem. biophys. Res. Comm.*, **65**, 233–240.
179. Jain, M.K. (1974), *Arch. Biochem. Biophys.*, **164**, 20–29.
180. Romine, W.O., Goodall, M.C., Peterson, J. and Bradley, R.J. (1974), *Biochim. biophys. Acta.*, **367**, 316–325.
181. Michaelson, D.M. and Raftery, M.A. (1974), *Proc. natn. Acad. Sci. U.S.A.*, **71**, 4768–4772.
182. Shamoo, A.E. and Eldefrawi, M.E. (1975), *J. Memb. Biol.*, **25**, 47–63.
183. McNamee, M.G., Weill, C.L. and Karlin, A. (1975), *Ann. N.Y. Acad. Sci.*, **264**, 175–182.
184. Howell, J., Kemp, G. and Eldefrawi, M.E., In: *Ion Transport Across Membranes* Proc. Joint U.S.-U.S.S.R. Conference (D.C. Tosteson, Y.A. Ovchinnikov and R. LaTorre, eds.) Raven Press, New York (in press).
185. Albuquerque, E.X., Barnard, E.A., Chiu, T.H., Lapa, A.J. Dolly, J.O., **70**, 949–953.
186. Daly, J.W., Karle, I., Myers, C.W., Tokuyama, T., Waters, J.A. and Witkop, B. (1971), *Proc. natn. Acad. Sci. U.S.A.*, **68**, 1870–1875.
187. Albuquerque, E.X., Kuba, K., Lapa, A.J., Daly, J.W. and Witkop, B. (1973), In: *Exploratory Concepts in Muscular Dystrophy II*, Proc. Int. Conf. of the Muscular Dystrophy Assoc. Oct. 14–19, *Excerpta Medica Intr. Congr.* Ser. No. 333, 585–597.

188. Albuquerque, E.X., Kuba, K. and Daly, J. (1974), *J. Pharmac. exp. Ther.*, **189**, 513–524.
189. Lapa, A.J., Albuquerque, E.X., Sarvey, J.M., Daly, J. and Witkop, B. (1975), *Exp. Neurol.*, **47**, 558–580.
190. Dolly, J.O., Albuquerque, E.X., Sarvey, J.M., Mallick, B. and Barnard, E.A. (1977), *Mol. Pharmacol.* **13**, 1–14.
191. Kato, G. and Changeux, J.-P. (1976), *Mol. Pharmacol.*, **12**, 92–100.
192. Cohen, J.B., Weber, M. and Changeux, J.-P. (1974), *Mol. Pharmacol.*, **10**, 35–40.
193. Eldefrawi, A.T., Eldefrawi, M.E., Albuquerque, E.X., Oliviera, A.C., Mansour, N., Adler, M., Daly, J., Brown, G.B., Burgermeister, W. and Witkop, B. *Proc. natn. Acad. Sci. U.S.A.*, (in press).
194. Kuo, J.F., Lee, T.P., Reyes, P.L., Walton, K.G., Donnelly, T.E., Jr. and Greengard, P. (1972), *J. biol. Chem.*, **247**, 16–22.
195. Lee, T.P., Kuo, J.F. and Greengard, P. (1972), *Proc. natn. Acad. Sci. U.S.A.*, **69**, 3287–3291.
196. Ferrendelli, J.A., Steiner, A.L., McDougal, D.R. and Kipnis, D.M. (1970), *Biochem. biophys. Res. Comm.*, **41**, 1061–1067.
197. Greengard, P. and Kebabian, J.W. (1974), *Fed. Proc.*, **33**, 1059–1067.
198. Kebabian, J.W., Bloom, F.E., Steiner, A.L. and Greengard, P. (1975), *Science*, **190**, 157–159.
199. Schultz, G., Hardman, J.G., Schultz, K., Davis, J.W. and Sutherland, E.W. (1973), *Proc. natn. Acad. Sci. U.S.A.*, **70**, 1721–1725.
200. Casnellie, J.E. and Greengard, P. (1974), *Proc. natn. Acad. Sci. U.S.A.*, **71**, 1891–1895.
201. Matsuzawa, H. and Nirenberg, M. (1975), *Proc. natn. Acad. Sci. U.S.A.*, **72**, 3472–3476.
202. Greengard, P. (1976), *Nature*, **260**, 101–108.
203. Schacht, J. and Agranoff, B.W. (1972), *J. biol. Chem.*, **247**, 771–777.
204. Schacht, J., Neale, E. and Agranoff, B.W. (1974), *J. Neurochem.*, **23**, 211–218.
205. Widlund, L. and Heilbronn, E. (1974), *J. Neurochem.*, **22**, 991–998.
206. Larrabee, M.G., Klingman, J.D. and Leicht, W.S. (1963), *J. Neurochem.*, **10**, 549–570.
207. Larrabee, M.G. and Leicht, W.S. (1965), *J. Neurochem.*, **12**, 1–13.
208. Rosenberg, P. (1973), *J. Pharm. Sci.*, **62**, 1552–1554.
209. Jafferji, S.S. and Michell, R.H. (1976), *Biochem. J.*, **154**, 653–658.
210. Yaghihara, Y., Bleasdale, J.E. and Hawthorne, J.N. (1973), *J. Neurochem.*, **21**, 173–190.
211. Katz, B. and Thesleff, S. (1957), *J. Physiol.*, **138**, 63–80.
212. Thesleff, S. (1959), *J. Physiol.*, **148**, 659–664.
213. Thesleff, S. (1955), *Acta Physiol. Scand.*, **34**, 218–231.
214. Nastuk, W.L. and Giessen, A.J. (1966), In: *Nerve as a Tissue* (K. Rodahl and B. Issekutz, Jr., eds.) Harper and Row Publ. New York, pp. 305–320.

215. Rang, H.P. and Ritter, J.M. (1970), *Mol. Pharmacol.*, **6**, 357–382.
216. Burgen, A.S.V. and Spero, L. (1968), *Br. J. Pharmac.*, **34**, 99–115.
217. Bennett, M.V.L., Wurzel, M. and Grundfest, H. (1961), *J. gen. Physiol.*, **44**, 757–804.
218. Lester, H.A., Changeux, J.-P. and Sheridan, R.E. (1975), *J. gen. Physiol.*, **65**, 797–816.
219. Paton, W.D.M. and Rothschild, A.M. (1965), *Br. J. Pharmac.*, **24**, 437–448.
220. Magazanik, L.G. and Vyskocil, F. (1970), *J. Physiol.*, **201**, 507–518.
221. Magazanik, L.G. and Vyskocil, F. (1975), *J. Physiol.*, **249**, 285–300.
222. Rang, H.P. and Ritter, J.M. (1970), *Mol. Pharmacol.*, **6**, 383–390.
223. Sugiyama, H., Popot, J.L., Cohen, J.B., Weber, M. and Changeux, J.-P. (1975), In: *Protein Ligand Interactions.* (H. Sund and G. Blauer, eds.) W. de Gruyter, New York, pp. 289–305.
224. Weber, M., David-Pfeuty, T. and Changeux, J.-P. (1975), *Proc. natn. Acad. Sci. U.S.A.*, **72**, 3443–3447.
225. Miledi, R. and Potter, L.T. (1971), *Nature*, **233**, 599–603.
226. Lester, H.A. (1972), *Mol. Pharmacol.*, **8**, 632–644.
227. Magazanik, L.G. and Vyskocil, F. (1973), In: *Drug Receptors* (H.P. Rang, ed.) Macmillan, London, pp. 105–119.
228. Popot, J.-L., Sugiyama, H. and Changeux, J.-P. (1974), *C.R. Acad. Sci. Ser. D.*, **279**, 1721–1724.
229. Cohen, J.B., Weber, M. and Changeux, J.-P. (1974), *Mol. Pharmacol.*, **10**, 904–932.
230. Birdsall, N.J.M. and Hulme, E.C. (1976), *J. Neurochem.*, **27**, 7–16.
231. Young, J.M. (1974), *FEBS Letters*, **46**, 354–356.
232. Eldefrawi, M.E. and O'Brien, R.D. (1971), *Proc. natn. Acad. Sci. U.S.A.*, **68**, 2006–2007.
233. Diamond, J. and Miledi, R. (1962), *J. Physiol.*, **162**, 393–408.
234. Axelsson, J. and Thesleff, S. (1959), *J. Physiol.*, **147**, 178–193.
235. Miledi, R. (1960), *J. Physiol.*, **151**, 1–23.
236. Vogel, Z., Sytkowski, A.J. and Nirenberg, M.W. (1972), *Proc. natn. Acad. Sci. U.S.A.*, **69**, 3180–3184.
237. Fambrough, D. and Rash, J.E. (1971), *Dev. Biol.*, **26**, 55–68.
238. Fischbach, G.D., Nameroff, M. and Nelson, P.G. (1971), *J. Cell Physiol.*, **78**, 289–300.
239. Patrick, J., Heinemann, S.F., Lindstrom, J., Schubert, D. and Steinbach, J.H. (1972), *Proc. natn. Acad. Sci. U.S.A.*, **69**, 2762–2766.
240. Sytkowski, A.J., Vogel, Z. and Nirenberg, M.W. (1973), *Proc. natn. Acad. Sci. U.S.A.*, **70**, 270–274.
241. Steinbach, J.H., Harris, A.J., Patrick, J., Schubert, D., Heinemann, S.F. (1973), *J. gen. Physiol.*, **62**, 255–270.
242. Miledi, R. and Zelena, J. (1966), *Nature*, **210**, 855–856.
243. Albuquerque, E.X. and McIsaac, R.J. (1970), *Exp. Neurol.*, **26**, 183–202.

244. Vogel, Z. and Daniels, M.P. (1976), *J. Cell Biol.*, **69**, 501–507.
245. Prives, J., Silman, I. and Amsterdam, A. (1976), *Cell*, **7**, 543–550.
246. Harris, A.J. and Dennis, M.J. (1970), *Science*, **167**, 1253–1255.
247. Cohen, S. and Fischbach, G. (1971), *Soc. for Neuroscience First Annual Meeting Abst.*, p. 162.
248. Greene, L.A., Sytkowski, A.J., Vogel, Z. and Nirenberg, M.W. (1973), *Nature*, **243**, 163–166.
249. Bekoff, A. and Betz, W.J. (1976), *Nature*, **193**, 915–917.
250. Vogel, Z. and Nirenberg, M. (1976), *Proc. natn. Acad. Sci. U.S.A.*, **73**, 1806–1810.
251. Enna, S.J., Yamamura, H.I. and Snyder, S. (1976), *Brain Res.*, **101**, 177–183.
252. Sastre, A., Gray, D.B. and Lane, M.-A. (1977), *Dev. Biol.*, **55**, 104–108.
253. Albuquerque, E.X., and Thesleff, S. (1968), *Acta physiol. scand.*, **73**, 471–480.
254. Albuquerque, E.X. and McIsaac, R.J. (1969), *Life Sci.*, **8**, 409–416.
255. Kuffler, S.W., Dennis, M.J. and Harris, A.J. (1971), *Proc. Roy. Soc. Ser. B.*, **177**, 555–563.
256. Dun, N., Mishi, S. and Karczmar, A.G. (1976), *Neuropharmacol.*, **15**, 211–218.
257. Berg, D.K. and Hall, Z.W. (1975), *J. Physiol.*, **244**, 659–676.
258. Berg, D.K., Kelly, R.B., Sargent, P.B., Williamson, P. and Hall, Z.W. (1972), *Proc. natn. Acad. Sci. U.S.A.*, **69**, 147–151.
259. Colquhoun, D., Rang, H.P. and Ritchie, J.M. (1974), *J. Physiol.*, **240**, 199–226.
260. Brockes, J.P. and Hall, Z.W. (1975), *Biochemistry*, **14**, 2100–2106.
261. Neher, E. and Sakmann, B. (1976), *J. Physiol.*, **258**, 705–729.
262. Trotter, J.L., Ringel, S.P., Cook, J.D., Engel, W.K., Eldefrawi, M.E. and McFarlin, D.E. *Neurology* (in press).
263. Guth, L. (1969), *Neurosci. Res. Prog. Bull.*, **7**, 3–73.
264. Murray, R.G. (1971), In: *Handbook of Sensory Physiology*, Vol. 4: *Chemical Senses*, Part 2: *Taste*. (L.M. Beidler, ed.), Springer Verlag, Berlin, pp. 31–50.
265. Albuquerque, E.X., Schuh, F.T. and Kauffman, F.C. (1971), *Pflügers Arch. Ges. Physiol.*, **328**, 123–137.
266. Deshpande, S.S., Albuquerque, E.X. and Guth, L. (1976), *Exp. Neurol.*, **53**, 151–165.
267. Albuquerque, E.X., Warnick, J.E., Tasse, J.R. and Sansone, F.M. (1972), *Exp. Neurol.*, **37**, 607–634.
268. Lavoie, P.-A., Collier, B. and Tenenhouse, A. (1976), *Science*, **260**, 349–350.
269. Drachman, D.B. and Johnston, D.M. (1975), *J. Physiol.*, **252**, 657–667.
270. Shainberg, A. and Burstein, M. (1976), *Nature*, **264**, 368–369.
271. Lomo, T. and Rosenthal, J. (1972), *J. Physiol.*, **221**, 493–513.
272. Drachman, D.B. and Witzke, F. (1972), *Science*, **176**, 514–516.
273. Pestronk, A., Drachman, D.B. and Griffin, J.W. (1976), *Science*, **260**, 352–353.

274. Giacobini, G., Filogamo, G., Weber, M., Boquet, P. and Changeux, J.-P. (1973), *Proc. natn. Acad. Sci. U.S.A.*, **70**, 1708–1712.
275. Crain, S.M. (1970), *J. exp. Zool.*, **173**, 353–369.
276. Crain, S.E. and Peterson, E.R. (1974), *Ann. N.Y. Acad. Sci.*, **228**, 6–34.
277. Kano, M. and Shimada, Y. (1971), *J. Cell Physiol.*, **78**, 233–242.
278. Robbins, N. and Yonezawa, T. (1971), *J. gen. Physiol.*, **58**, 467–481.
279. Kidokoro, Y. and Heinemann, S. (1974), *Nature*, **252**, 593–594.
280. Nurse, C.A. and O'Lague, P.H. (1975), *Proc. natn. Acad. Sci. U.S.A.*, **72**, 1955–1959.
281. Puro, D.G. and Nirenberg, M. (1976), *Proc. natn. Acad. Sci. U.S.A.*, **73**, 3544–3548.
282. Chang, C.C. and Tung, L.H. (1974), *Eur. J. Pharmacol.*, **26**, 386–388.
283. Fambrough, D.M. (1970), *Science*, **168**, 372–373.
284. Grampp, W., Harris, J.B. and Thesleff, S. (1972), *J. Physiol.*, **221**, 743–754.
285. Brockes, J.P. and Hall, Z.W. (1975), *Proc. natn. Acad. Sci. U.S.A.*, **72**, 1368–1372.
286. Merlie, J.P., Sobel, A., Changeux, J.-P. and Gros, F. (1975), *Proc. natn. Acad. Sci. U.S.A.*, **72**, 4028–4032.
287. Merlie, J.P., Changeux, J.-P. and Gros, F. (1976), *Science*, **264**, 74–76.
288. Devreotes, P.N. and Fambrough, D.M. (1976), *Proc. natn. Acad. Sci. U.S.A.*, **73**, 161–164.
289. Devreotes, P.N. and Fambrough, D.M. (1975), *J. Cell Biol.*, **65**, 335–358.
290. Berg. D.K. and Hall, Z.W. (1975), *J. Physiol.*, **252**, 771–789.
291. Berg, D.K. and Hall, Z.W. (1974), *Science*, **184**, 473–475.
292. Simpson, J.A. (1960), *Scot. Med. J.*, **5**, 419 436.
293. Albuquerque, E.X., Rash, J.E., Mayer, R.F. and Satterfield, J.R. (1976), *Exp. Neurol.*, **51**, 536–563.
294. Bender, A.N., Ringel, S.P., Engel, W.K., Daniels, M.P. and Vogel, Z. (1975), *Lancet*, **1**, 607–609.
295. Almon, R.R. and Appel, S.H. (1976), *Ann. N.Y. Acad. Sci.*, **274**, 235–243.
296. Almon, R.R., Andrew, C.G. and Appel, S.N. (1974), *Science*, **186**, 55–57.
297. Appel, S., Almon, R. and Levy, N. (1975), *N. Engl. J. Med.*, **293**, 760–761.
298. Abramsky, O., Aharonov, A., Webb, C. and Fuchs, S. (1975), *Clin. exp. Immunol.*, **19**, 11–16.
299. Patrick, J. and Lindstrom, J. (1973), *Science*, **180**, 871–872.
300. Heilbronn, E., Mattsson, C., Thornell, L.-E., Sjöström, M., Stalberg, E., Hilton-Brown, P. and Elmqvist, D. (1976), *Ann. N.Y. Acad. Sci.*, **274**, 337–353.
301. Sugiyama, H., Benda, P., Meunier, J.-C. and Changeux, J.-P. (1973), *FEBS Letters*, **35**, 124–128.
302. Sanders, D.B., Schleifer, L.S., Eldefrawi, M.E., Norcross, N.L. and Cobb, E.E. (1976), *Ann. N.Y. Acad. Sci.*, **274**, 319–336.
303. Penn, A.S., Chang, H.W., Lovelace, R.E., Niemi, W. and Miranda, A. (1976), *Ann. N.Y. Acad. Sci.*, **274**, 354–376.

304. Lennon, V.A., Lindstrom, J.M. and Seybold, M.E. (1975), *J. exp. Med.*, **141**, 1365–1375.

305. Seybold, M.E., Lambert, E.H., Lennon, V.A. and Lindstrom, J.M. (1976), *Ann. N.Y. Acad. Sci.*, **274**, 275–282.

306. Sanders, D.B., Johns, T.R., Eldefrawi, M.E. and Cobb, E.E. (1977), *Arch. Neurol.*, **34**, 75–79.

307. Toyka, K.V., Drachman, D.B., Griffin, D.E., Pestronk, A., Winkelstein, J.A., Fischbeck, K.H., Jr. and Kao, I. (1977), *New Engl. J. Med.*, **296**, 125–131.

308. Bevan, S., Heinemann, S., Lennon, V.A. and Lindstrom, J. (1976), *Science*, **260**, 438–439.

309. Shoulson, I. and Chase, T.N. (1975), *Ann. Rev. Med.*, **26**, 419–436.

310. Mones, R.J. (1973), In: *Advances in Neurology.* (A. Barbeau, T.N. Chase and G.W. Paulson, eds.), Raven Press, New York, pp. 665–669.

311. Dill, R.E., Dorris, R.L. and Phillips-Thonnard, I. (1976), *J. Pharm. Pharmac.*, **28**, 646–648.

312. Enna, S.J., Bird, E.D., Bennett, J.P., Jr., Byland, D.B., Yamamura, H. I., Iverson, L.L. and Snyder, S.H. (1976), *N. Eng. J. Med.*, **294**, 1305–1309.

313. Hiley, R. (1976), In: *Biochemistry and Neurology.* (H.F. Bradford and C.D. Marsden, eds.), Academic Press, New York, pp. 103–109.

314. Bloom, F.E., Costa, E. and Salmoiraghi, G.C. (1965), *J. Pharmac. exp. Ther.*, **150**, 244–252.

315. McLennan, H. and York, D.H. (1966), *J. Physiol.*, **187**, 163–175.

316. Hiley, C.R. and Bird, E.D. (1974), *Brain Res.*, **80**, 355–358.

317. Till, J.E. and McCulloch, E.A. (1961), *Rad. Res.*, **14**, 213–222.

318. Becker, A.J., McCulloch, E.A., Siminovitch, L. and Till, J.E. (1965), *Blood*, **26**, 296–308.

319. Byron, J.W. (1974), In: *Cell Cycle Controls.* (G.M. Padilla , I.L. Cameron and A. Zimmerman, eds), Academic Press, New York, pp. 87–99.